THE LAST THREE MINUTES

THE LAST
THREE MINUTES

Conjectures about the Ultimate Fate of the Universe

PAUL DAVIES

BasicBooks

A Division of HarperCollins*Publishers*

The Science Masters Series is a global publishing venture consisting of original science books written by leading scientists and published by a worldwide team of twenty-six publishers assembled by John Brockman. The series was conceived by Anthony Cheetham of Orion Publishers and John Brockman of Brockman Inc., a New York literary agency, and developed in coordination with BasicBooks.

The Science Masters name and marks are owned by and licensed to the publisher by Brockman Inc.

•••••

Designed by Joan Greenfield

•••••

LIBRARY OF CONGRESS CATALOGING-IN-PUBLICATION DATA
Davies, P. C. W.
    The last three minutes : conjectures about the ultimate fate of the universe / by Paul Davies.
        p.   cm. — (Science masters series)
    Includes bibliographical references and index.
    ISBN 0-465-04892-7
    1. Cosmology—Popular works. I. Title. II. Series.
QB982.D38 1994
523.1'9—dc20   94-6345
                                                    CIP

•••••

94 95 96 97 ❖/RRD 9 8 7 6 5 4 3 2 1

．．

And so some day,

The mighty ramparts of the mighty universe

Ringed round with hostile force,

Will yield and face decay and come crumbling to ruin.

—Lucretius, *De Rerum Natura*

CONTENTS

When I was a student in the early 1960s, there was intense interest in the problem of the origin of the universe. The big-bang theory, which dated from the 1920s but had been considered seriously only since the 1950s, was well known but far from convincing. The rival steady-state theory, which did away with a cosmic origin altogether, was still the most fashionable scenario in some quarters. Then came the discovery of the cosmic background heat radiation by Robert Penzias and Arno Wilson in 1965, and the subject was transformed. This, surely, was clear evidence for a hot, violent, abrupt origin of the universe.

Cosmologists feverishly worked through the implications of the discovery. How hot was the universe one million years after the big bang? One year after? One second? What sort of physical processes would have occurred in that primeval inferno? Might there be relics from the dawn of creation retaining an imprint of the extreme conditions that must then have prevailed?

I well remember attending a lecture on cosmology in 1968. The professor finished by reviewing the big-bang theory in the light of the discovery of the background heat radiation. "Some theoreticians have given an account of the chemical composition of the universe based on the nuclear processes that occurred during the first three minutes after the big bang," he related with a smile. Everyone in the audience laughed uproariously. It seemed absurdly ambitious to try to describe the state of the universe just moments after it had come into existence. Even the seventeenth-century archbishop James Ussher, whose study of the minutiae of biblical chronology led him to declare that the universe was created on October 23, 4004 B.C., did not have the temerity to catalog the precise sequence of events during the first three minutes.

Such is the pace of scientific progress that barely a decade after the discovery of the cosmic background heat radiation, the first three minutes had become standard fare for students. Textbooks were written on the subject. Then in 1977 the American physicist and cosmologist Steven Weinberg published a best-selling book aptly titled *The First Three Minutes*. It proved to be a landmark in popular science publishing. Here was one of the world's experts providing the general public with a detailed and totally convincing account of processes that occurred mere moments after the big bang.

While the public was catching up with these heady developments, the scientists themselves were moving on. Attention began to shift from what had become known as the early universe—meaning minutes or so after the big bang—to the *very* early universe—an almost infinitesimal fraction of a second after the beginning. Another decade or so later and the British mathematical physicist Stephen Hawking was confidently able to describe, in *A Brief History of Time,* the latest ideas about the first trillion-trillion-trillionth of a second. The laughter ending that lecture in 1968 now seems rather hollow.

With the big-bang theory well established in both the popular and the scientific mind, more and more thought is being given to the future of the universe. We have a good idea how the universe began, but how will it end? What can we say about its ultimate fate? Will the universe finish with a bang or a whimper—indeed, will it ever end at all? And what of us? Can humanity or our descendants, be they robotic or flesh and blood, survive for all eternity?

It is impossible not to be curious about such matters, even though Armageddon may not be just around the corner. Our struggle for survival on planet Earth, which is currently beset by man-made crises, is placed in a welcome new context when we are forced to reflect on the cosmological dimension of our existence. *The Last Three Minutes* is the story of the future of the universe, as best we can predict it, based on the latest thinking by some well-known physicists and cosmologists. It is not all apocalyptic. In fact, the future holds the promise of unprecedented potential for development and richness of experience. But we cannot ignore the fact that what can come to exist can also cease to exist.

This book is intended for the general reader. No prior knowledge of science or mathematics is necessary. From time to time, however, I need to discuss very large or very small numbers, and it helps to use a compact mathematical notation known as the powers of ten to represent them. The number one hundred billion, for example, when written out in full, is 100,000,000,000, which is rather cumbersome. There are eleven zeros after the 1 in this number, so we can represent it by writing 10^{11}—in words, "ten to the eleventh power." Similarly, one million is 10^6 and one trillion is 10^{12}. And so on. Remember, however, that this notation tends to camouflage the rate at which these numbers grow: 10^{12} is a hundred times bigger than 10^{10}; it is a *much* bigger number, even though it looks almost the same. Powers of ten, used negatively, can also represent very small numbers: thus the fraction one-billionth, or 1/1,000,000,000, becomes 10^{-9} ("ten to the minus nine"), because there are nine zeros after the 1 on the bottom of the fraction.

Finally, I should like to warn the reader that this book is necessarily highly speculative. While most of the ideas to be presented are based on our best current understanding of science, futurology cannot enjoy the same status as other scientific endeavors. Nevertheless, the temptation to speculate about the ultimate destiny of the cosmos is irresistible. It is in this spirit of open-minded inquiry that I have written this book. The basic scenario of a universe originating in a big bang, then expanding and cooling toward some final state of physical degeneration, or perhaps collapsing catastrophically, is fairly well established scientifically. What is far less certain, however, are the dominant physical processes that may occur over the immense time scales involved. Astronomers have a clear idea of the general fate of ordinary stars, and are becoming increasingly confident that they understand the basic properties of neutron stars and black holes; but if the universe endures for many trillions of years or more, there may be subtle physical effects about which we can only conjecture that eventually become very important.

When faced with the problem of our incomplete understanding of nature, all we can do to try and deduce the ultimate fate of the universe is to employ our best existing theories and extrapolate them to their logical conclusions. The problem is that many of the theories having an important bearing on the fate of the universe remain to be tested experimentally. Some of the processes I discuss—for example, gravitational-wave emission, proton decay, and black-hole radiance—are enthusiastically believed by theorists but have not yet been observed. Just as seriously, there will undoubtedly be other physical processes we know nothing about that could dramatically alter the ideas presented here.

These uncertainties become even greater when we consider the possible effects of intelligent life in the universe. Here we enter the realm of science fiction; nevertheless, we cannot ignore the fact that living beings may, over the eons, significantly modify the behavior of physical systems on ever-larger scales. I decided to include the topic of life in the cosmos because for many readers the fascination with the fate of the universe is intimately bound up with their concern for the fate of human beings, or the remote descendants of human beings. We should remember, though, that scientists have no real understanding of the nature of human consciousness, nor of the physical requirements that may permit conscious activity to continue in the far future of the universe.

I should like to thank John Barrow, Frank Tipler, Jason Twamley, Roger Penrose, and Duncan Steel for helpful discussions about the subject matter of this book; the series editor, Jerry Lyons, for his critical reading of the manuscript; and Sara Lippincott, for her excellent work on the final manuscript.

The date: August 21, 2126. Doomsday.

The place: Earth. Across the planet a despairing population attempts to hide. For billions there is nowhere to go.
Some people flee deep underground, desperately seeking out
caves and disused mine shafts, or take to the sea in sub-
marines. Others go on the rampage, murderous and uncar-
ing. Most just sit, sullen and bemused, waiting for the end.

High in the sky, a huge shaft of light is etched into the fab-
ric of the heavens. What began as a slender pencil of softly
radiating nebulosity has swollen day by day to form a mael-
strom of gas boiling into the vacuum of space. At the apex of
a vapor trail lies a dark, misshapen, menacing lump. The
diminutive head of the comet belies its enormous destructive
power. It is closing on planet Earth at a staggering 40,000
miles per hour, 10 miles every second—a trillion tons of ice
and rock, destined to strike at seventy times the speed of
sound.

Mankind can only watch and wait. The scientists, who
have long since abandoned their telescopes in the face of the
inevitable, quietly shut down the computers. The endless
simulations of disaster are still too uncertain, and their conclusions are too alarming to release to the public anyway.

Some scientists have prepared elaborate survival strategies, using their technical knowledge to gain advantage over their fellow citizens. Others plan to observe the cataclysm as carefully as possible, maintaining their role as true scientists to the very end, transmitting data to time capsules buried deep in the Earth. For posterity. . . .

The moment of impact approaches. All over the world, millions of people nervously check their watches. The last three minutes.

Directly above ground zero, the sky splits open. A thousand cubic miles of air are blasted aside. A finger of searing flame wider than a city arcs groundward and fifteen seconds later lances the Earth. The planet shudders with the force of ten thousand earthquakes. A shock wave of displaced air sweeps over the surface of the globe, flattening all structures, pulverizing everything in its path. The flat terrain around the impact site rises in a ring of liquid mountains several miles high, exposing the bowels of the Earth in a crater a hundred miles across. The wall of molten rock ripples outward, tossing the landscape about like a blanket flicked in slow motion.

Within the crater itself, trillions of tons of rock are vaporized. Much more is splashed aloft, some of it flung out into space. Still more is pitched across half a continent to rain down hundreds or even thousands of miles away, wreaking massive destruction on all beneath. Some of the molten ejecta falls into the ocean, raising huge tsunamis that add to the spreading turmoil. A vast column of dusty debris fans out into the atmosphere, blotting out the sun across the whole planet. Now the sunlight is replaced by the sinister, flickering glare of a billion meteors, roasting the ground below with their searing heat, as displaced material plunges back from space into the atmosphere.

The preceding scenario is based on the prediction that comet Swift-Tuttle will hit the earth on August 21, 2126. If it were to, global devastation would undoubtedly follow, destroying human civilization. When this comet paid us a visit in 1993, early calculations suggested that a collision in 2126 was a distinct possibility. Since then, revised calculations indicate that the comet will in fact miss Earth by two weeks: a close shave, but we can breathe easily. However, the danger won't go away entirely. Sooner or later Swift-Tuttle, or an object like it, *will* hit the Earth. Estimates suggest that 10,000 objects half a kilometer or more in diameter move on Earth-intersecting orbits. These astromomical interlopers originate in the frigid outer reaches of the solar system. Some are the remains of comets that have become trapped by the gravitational fields of the planets, others come from the asteroid belt that lies between Mars and Jupiter. Orbital instability causes a continual traffic of these small but lethal bodies into and out of the inner solar system, constituting an ever-present menace to Earth and our sister planets.

Many of these objects are capable of causing more damage than all the world's nuclear weapons put together. It is only a matter of time before one strikes. When it does, it will be bad news for people. There will be an abrupt and unprecedented interruption in the history of our species. But for the Earth such an event is more or less routine. Cometary or asteroid impacts of this magnitude occur, on average, every few million years. It is widely believed that one or more such events caused the extinction of the dinosaurs sixty-five million years ago. It could be us next time.

Belief in Armageddon is deep-rooted in most religions and cultures. The biblical book of Revelation gives a vivid account of the death and destruction that lie in store for us:

> Then there came flashes of lightning, rumblings, peals of thunder, and a severe earthquake. No earthquake like it has ever occurred since man has been on Earth, so tremendous was the quake. . . . The cities of the

nations collapsed. . . . Every island fled away and the mountains could not be found. From the sky huge hailstones of about a hundred pounds each fell upon men. And they cursed God on account of the plague of hail, because the plague was so terrible.

There are certainly lots of nasty things that could happen to Earth, a puny object in a universe pervaded by violent forces, yet our planet has remained hospitable to life for at least three and a half billion years. The secret of our success on planet Earth is space. Lots of it. Our solar system is a tiny island of activity in an ocean of emptiness. The *nearest* star (after the sun) lies more than four light-years away. To get some idea of how far that is, consider that light traverses the ninety-three million miles from the sun in only eight and a half minutes. In four years, it travels more than twenty trillion miles.

The sun is a typical dwarf star, lying in a typical region of our galaxy, the Milky Way. The galaxy contains about a hundred billion stars, ranging in mass from a few percent to a hundred times the mass of the sun. These objects, together with a lot of gas clouds and dust and an uncertain number of comets, asteroids, planets, and black holes, slowly orbit the galactic center. Such a huge collection of bodies may give the impression that the galaxy is a very crowded system, until account is taken of the fact that the visible part of the Milky Way measures about a hundred thousand light-years across. It is shaped like a plate, with a central bulge; a few spiral arms made up of stars and gas are strung out around it. Our sun is located in one such spiral arm and is about thirty thousand light-years from the middle.

As far as we know, there is nothing very exceptional about the Milky Way. A similar galaxy, called Andromeda, lies about two million light-years away, in the direction of the constellation of that name. It can just be seen with the unaided eye as a fuzzy patch of light. Many billions of galaxies, some spiral, some elliptical, some irregular, adorn the observable universe. The scale of distance is vast. Powerful telescopes can image individual galaxies several *billion* light-years away. In some cases, it has taken their light longer than the age of the Earth (four and a half billion years) to reach us.

All this space means that cosmic collisions are rare. The greatest threat to Earth is probably from our own backyard. Asteroids do not normally orbit close to Earth; they are largely confined to the belt between Mars and Jupiter. But the huge mass of Jupiter can disturb the aster- oids' orbits, occasionally sending one of them plunging in toward the sun, and thus menacing Earth.

Comets pose another threat. These spectacular bodies are believed to originate in an invisible cloud situated about a light year from the sun. Here the threat comes not from Jupiter but from passing stars. The galaxy is not sta- tic; it rotates slowly, as its stars orbit the galactic nucleus. The sun and its little retinue of planets take about two hundred million years to complete one circuit of the galaxy, and on the way they have many adventures. Nearby stars may brush the cloud of comets, displacing a few toward the sun. As the comets plunge through the inner solar system, the sun evaporates some of their volatile material, and the solar wind blows it out in a long streamer—the famous cometary tail. Very rarely, a comet will collide with the Earth during its sojourn in the inner solar system. The comet does the damage, but the passing star must bear the responsibility. Fortunately, the huge distances between the stars insulate us against too many such encounters.

Other objects can also pass our way on their journey around the galaxy. Giant clouds of gas drift slowly by, and though they are more tenuous even than a laboratory vac- uum they can drastically alter the solar wind and may affect the heat flow from the sun. Other, more sinister objects may lurk in the inky depths of space: rogue planets, neutron stars, brown dwarfs, black holes—all these and more could come upon us unseen, without warning, and wreak havoc with the solar system.

Or the threat could be more insidious. Some astronomers believe that the sun may belong to a double-star system, in common with a great many other stars in the galaxy. If it exists, our companion star—dubbed Nemesis, or the Death Star—is too dim and too far away to have been discovered yet. But in its slow orbit around the sun it could still make its presence felt gravitationally, by periodically disturbing distant comets and sending some plunging Earthward to produce a series of devastating impacts. Geologists have found that wholesale ecological destruction does indeed occur periodically—about every thirty million years.

Looking farther afield, astronomers have observed entire galaxies in apparent collision. What chance is there that the Milky Way will be smashed by another galaxy? There is some evidence, in the very rapid movement of certain stars, that the Milky Way may have already been disrupted by collisions with small nearby galaxies. However, the collision of two galaxies does not necessarily spell disaster for their constituent stars. Galaxies are so sparsely populated that they can merge into one another without individual stellar collisions.

Most people are fascinated by the prospect of Doomsday—the sudden, spectacular destruction of the world. But violent death is less of a threat than slow decay. There are many ways in which Earth could gradually become inhospitable. Slow ecological degradation, climatic change, a small variation in the heat output of the sun—all these could threaten our comfort, if not survival, on our fragile planet. Such changes, however, will take place over thousands or even millions of years, and humanity may be able to combat them using advanced technology. The gradual onset of a new ice age, for example, would not spell total disaster for our species, given the time available to reorganize our activities. One can speculate that technology will continue to advance dramatically over the coming millennia; if so, it is tempting to believe that human beings, or their descendants, will gain control over ever-larger physical systems and may eventually be in a position to avert disasters even on an astronomical scale.

Can humanity, in principle, survive forever? Possibly. But we shall see that immortality does not come easily and may yet prove to be impossible. The universe itself is subject to physical laws that impose upon it a life cycle of its own: birth, evolution, and—perhaps—death. Our own fate is entangled inextricably with the fate of the stars.

....

CHAPTER 2

THE DYING UNIVERSE

In the year 1856, the German physicist Hermann von Helmholtz made what is probably the most depressing prediction in the history of science. The universe, Helmholtz claimed, is dying. The basis of this apocalyptic pronouncement was the so-called second law of thermodynamics. Originally formulated in the early nineteenth century as a rather technical statement about the efficiency of heat engines, the second law of thermodynamics (now often termed simply "the second law") was soon recognized as having universal significance—indeed, literally cosmic consequences.

In its simplest version, the second law states that heat flows from hot to cold. This is, of course, a familiar and obvious property of physical systems. We see it at work whenever we cook a meal or let a hot cup of coffee cool: the heat flows from the region with the higher temperature to that with the lower temperature. There is no mystery about this. Heat manifests itself in matter in the form of molecular agitation. In a gas, such as air, the molecules rush around chaotically and collide. Even in a solid body the atoms jiggle vigorously about. The hotter the body, the more energetic the molecular agitation will be. If two bodies of different temperature are brought into contact, the

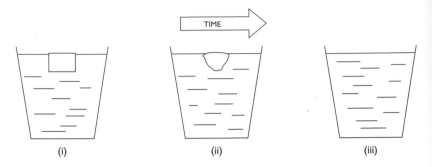

FIGURE 2.1

The arrow of time. The melting ice cube defines a directionality in time: heat flows from the warm water into the cold ice.
A movie showing the sequence (iii), (ii), (i) would soon be recognized as a trick. This asymmetry is characterized by a quantity called entropy, which rises as the ice melts.

• • • • • • • • •

more vigorous molecular agitation in the hot body soon
spreads to the molecules of the cooler body.

Because heat flow is unidirectional, the process is lop-
sided in time. A movie showing heat flowing spontaneously from cold to hot would look as silly as a river
flowing uphill or raindrops rising to the clouds. So we can
identify a fundamental directionality to heat flow, often
represented by an arrow pointing from past to future. This
"arrow of time" indicates the irreversible nature of thermodynamic processes and has fascinated physicists for a
hundred and fifty years. (See figure 2.1).

The work of Helmholtz, Rudolf Clausius, and Lord
Kelvin led to the recognition of the significance of a quantity called entropy for characterizing irreversible change
in thermodynamics. In the simple case of a hot body in
contact with a cold body, the entropy can be defined as
heat energy divided by temperature. Consider a small
quantity of heat flowing from the hot body into the cold
body. The hot body will lose some entropy and the cold

body will gain some. Because the same quantity of heat energy is involved but the temperatures differ, the entropy gained by the cold body will be greater than that lost by the hot body. Thus the total entropy of the whole system— hot body plus cold body—rises. One statement of the second law of thermodynamics is then that the entropy of such a system should never fall, for to do so would imply that some heat had gone spontaneously from cold to hot.

A more thoroughgoing analysis enables this law to be generalized to all closed systems: the entropy never falls. If the system includes a refrigerator, which *can* transfer heat from cold to hot, totaling the entropy of the whole system must take into account the energy expended in running the refrigerator. The process of expenditure will itself increase the entropy. It is then always the case that the entropy created by running the refrigerator more than offsets the reduction in entropy resulting from the transfer of heat from cold to hot. In natural systems, too, such as those involving biological organisms or the formation of crystals, the entropy of one part of the system often falls, but this fall is always paid for by a compensatory rise in entropy in another part of the system. Overall, the entropy never goes down.

If the universe as a whole can be considered as a closed system, on the basis that there is nothing "outside" it, then the second law of thermodynamics makes an important prediction: the total entropy of the universe never decreases. In fact, it goes on rising remorselessly. A good example lies right on our cosmic doorstep—the sun, which continuously pours heat into the cold depths of space. The heat goes off into the universe, never to return; this is a spectacularly irreversible process.

An obvious question is, Can the entropy of the universe go on rising forever? Imagine a hot body and a cold body brought into contact inside a thermally sealed container. Heat energy flows from hot to cold and the entropy rises, but eventually the cold body will warm up and the hot body will cool down so that they reach the same temperature. When that state is achieved, there will be no further heat transfer. The system inside the container will have reached a uniform temperature—a stable state of maximum entropy referred to as thermodynamic equilibrium. No further change is expected, as long as the system remains isolated; but if the bodies are disturbed in some way—say, by introducing more heat from outside the container—then further thermal activity will occur, and the entropy will rise to a higher maximum.

What do these basic thermodynamic ideas tell us about astronomical and cosmological change? In the case of the sun and most other stars, the outflow of heat can continue for many billions of years, but it is not inexhaustible. A normal star's heat is generated by nuclear processes in its interior. As we shall see, the sun will eventually run out of fuel, and unless overtaken by events it will cool until it reaches the same temperature as the surrounding space.

Although Hermann von Helmholtz knew nothing of nuclear reactions (the source of the sun's immense energy was a mystery at that time), he understood the general principle that all physical activity in the universe tends toward a final state of thermodynamic equilibrium, or maximum entropy, following which nothing of value is likely to happen for all eternity. This one-way slide toward equilibrium became known to the early thermodynamicists as the "heat death" of the universe. Individual systems, it was conceded, might be revitalized by external disturbances, but the universe itself had no "outside" by definition, so nothing could prevent an all-encompassing heat death. It seemed inescapable.

The discovery that the universe was dying as an inexorable consequence of the laws of thermodynamics had a profoundly depressing effect on generations of scientists and philosophers. Bertrand Russell, for example, was moved to write the following gloomy assessment in his book *Why I Am Not a Christian*:

> All the labours of the ages, all the devotion, all the inspiration, all the noonday brightness of human genius, are destined to extinction in the vast death of the solar system, and . . . the whole temple of man's achievement must inevitably be buried beneath the debris of a universe in ruins—all these things, if not quite beyond dispute, are yet so nearly certain that no philosophy which rejects them can hope to stand. Only within the scaffolding of these truths, only on the firm foundation of unyielding despair, can the soul's habitation henceforth be safely built.

Many other writers have concluded from the second law of thermodynamics and its implication of a dying universe that the universe is pointless and human existence ultimately futile. I shall return to this bleak assessment in later chapters and discuss whether or not it is misconceived.

The prediction of a final cosmic heat death not only says something about the future of the universe but also implies something important about the past. It is clear that if the universe is irreversibly running down at a finite rate, then it cannot have existed forever. The reason is simple: if the universe were infinitely old, it would have died already. Something that runs down at a finite rate obviously cannot have existed for eternity. In other words, the universe must have come into existence a finite time ago.

It is remarkable that this profound conclusion was not properly grasped by the scientists of the nineteenth century. The idea of the universe originating abruptly in a big bang had to await astronomical observations in the 1920s, but a definite genesis at some moment in the past seems to have been strongly suggested already, on purely thermodynamic grounds.

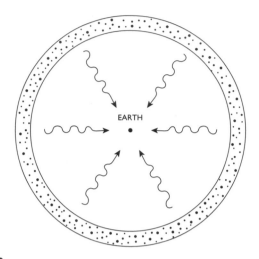

FIGURE 2.2

Olbers' paradox. Imagine an unchanging universe populated by randomly scattered stars at a uniform average density. Shown is a selection of stars occupying a thin spherical shell of space centered on Earth. (The stars outside the shell have been omitted from the picture.) Light from the stars in this shell contributes to the total flux of starlight falling on Earth. The intensity of light from a given star will diminish as the square of the shell's radius. However, the total number of stars in the shell will grow in proportion to the square of the shell's radius. Therefore these two factors cancel each other out, and the total luminosity of the shell is *independent* of its radius. In an infinite universe, there will be an infinity of shells and—apparently—an infinite flux of light reaching Earth.

· · · · · · · · ·

Because this obvious inference was not made, however, nineteenth-century astronomers were baffled by a curious cosmological paradox. Known as Olbers' paradox, after the German astronomer who is credited with its formulation, it poses a simple yet deeply significant question: Why is the sky dark at night?

At first, the problem seems trivial. The night sky is dark because the stars are situated at immense distances from us and so appear dim. (See figure 2.2.) But suppose that space has no limit. In this case, there could well be an infinity of stars. An infinite number of dim stars would add up to a lot of light. It is easy to calculate the cumulative starlight from an infinity of unchanging stars distributed more or less uniformly throughout space. The brightness of a star diminishes with distance, according to an inverse-square law. This means that at twice the distance the star is one-quarter as bright, at three times the distance it is one-ninth as bright, and so on. On the other hand, the number of stars increases the farther away you look. In fact, simple geometry shows that the number of stars, say, two hundred light-years away is four times the number one hundred light-years away, while the number three hundred light-years away is nine times the latter. So the number of stars goes up as the square of the distance, while the brightness goes down as the square of the distance. The two effects cancel each other out, and the result is that the total light coming from all the stars at a given distance does not depend on the distance. The same total light comes from stars two hundred light-years away as from those one hundred light-years away.

The problem comes when we add up the light from all the stars at all possible distances. If the universe has no boundary, there seems to be no limit to the total amount of light received on Earth. Far from being dark, the night sky ought to be infinitely bright!

The problem is ameliorated somewhat when account is taken of the finite size of stars. The farther away a star is from Earth, the smaller is its apparent size. A nearby star will obscure a more distant star if it lies along the same line of sight. In an infinite universe this will happen infinitely often, and taking it into account changes the conclusion of the previous calculation. Instead of an infinite flux of light arriving on Earth, the flux is merely very large—roughly equivalent to the sun's disk filling the sky, as would be the case if the Earth were located about a million miles from the solar surface. This would be a very uncomfortable location indeed; in fact, the Earth would be rapidly vaporized by the intense heat.

The conclusion that an infinite universe ought to be a cosmic furnace is actually a restatement of the thermodynamic problem I discussed earlier. The stars pour heat and light into space, and this radiation slowly accumulates in the void. If the stars have been burning forever, it seems at first sight that the radiation must have an infinite intensity. But some radiation, while traveling through space, will strike other stars and be reabsorbed. (This is equivalent to noticing that nearby stars obscure the light from more distant ones.) Therefore the intensity of the radiation will rise until an equilibrium is established at which the rate of emission just balances the rate of absorption. This state of thermodynamic equilibrium will occur when the radiation in space reaches the same temperature as the surfaces of the stars—a few thousand degrees. Thus the universe should be full of heat radiation with a temperature of several thousand degrees, and the night sky, instead of being dark, should glow at this temperature.

Heinrich Olbers proposed a resolution to his own paradox. Noting the existence of large amounts of dust in the universe, he suggested that this material would absorb most of the starlight and thus darken the sky. Unfortunately, his idea, though imaginative, was fundamentally flawed: the dust would eventually heat up and start to glow with the same intensity as the radiation it absorbed.

Another possible resolution is to abandon the assumption that the universe is infinite in extent. Suppose the stars are many but finite in number, so that the universe consists of a huge assemblage of stars surrounded by an infinite dark void; then most of the starlight will flow away into the space beyond, and be lost. But this simple resolution, too, has a fatal flaw—one that was, in fact, already familiar to Isaac Newton in the seventeenth century. The flaw concerns the nature of gravitation: Every star attracts every other star with a force of gravity, therefore all the stars in the assemblage would tend to fall together, congregating at the center of gravity. If the universe has a definite center and edge, it seems that it must collapse in on itself. An unsupported, finite, static universe is unstable, and subject to gravitational collapse.

This gravitational problem will crop up again later in my story. Here we need simply note the ingenious way in which Newton attempted to sidestep it. The universe can collapse to its center of gravity, Newton reasoned, only if it *has* a center of gravity. If the universe is both infinite in extent and (on average) uniformly populated with stars, then there will be no center and no edge. A given star will be pulled every which way by its many neighbors, like a gigantic tug-of-war in which ropes bristle in all directions. On average, all these tugs will cancel one another, and the star won't move.

So if we accept Newton's resolution of the collapsing-cosmos paradox, we are back with an infinite universe again, and the problem of Olbers' paradox. It seems that we must face either one or the other. But with the benefit of hindsight we can find a way between the horns of the dilemma. It is not the assumption that the universe is infinite in *space* that is wrong but the assumption that it is infinite in *time*. The paradox of the flaming sky arose because astronomers assumed that the universe was unchanging, that the stars were static and had been burning with undiminished intensity for all eternity. But we now know that both these assumptions were wrong. First, as I shall shortly explain, the universe is not static but expanding. Second, the stars cannot have been burning forever, because they would have long since run out of fuel. The fact that they are burning now implies that the universe must have come into existence at a finite time in the past.

If the universe has a finite age, Olbers' paradox goes away immediately. To see why, consider the case of a very distant star. Because light travels at a finite speed (300,000 kilometers a second, in a vacuum) we do not see the star as it is today but as it was when the light left it. For example, the bright star Betelgeuse is about six hundred and fifty light-years away, so it appears to us now as it was six hundred and fifty years ago. If the universe came into existence, say, ten billion years ago, then we would not see any stars located more than ten billion light-years away from Earth. The universe *may* be infinite in spatial extent, but if it has a finite age we cannot in any case see beyond a certain finite distance. So the cumulative starlight from an infinite number of stars of finite *age* will be finite, and possibly insignificantly small.

The same conclusion follows from thermodynamic considerations. The time taken for the stars to fill space with heat radiation and reach a common temperature is immense, because there is so much empty space in the universe. There has simply been insufficient time since the beginning for the universe to have reached thermodynamic equilibrium by now.

All the evidence points, then, to a universe that has a limited life span. It came into existence at some finite time in the past, it is currently vibrant with activity, but it is inevitably degenerating toward a heat death at some stage in the future. A host of questions immediately arises. When will the end come? What form will it take? Will it be slow or sudden? And is it conceivable that the heat-death conclusion, as scientists currently understand it, might turn out to be wrong?

CHAPTER 3

···

THE FIRST THREE MINUTES

Cosmologists, like historians, understand that the key to the future lies in the past. In the last chapter, I explained how the laws of thermodynamics suggest a universe of limited longevity. There is almost unanimous opinion among scientists that the entire cosmos originated between ten and twenty billion years ago in a big bang, and that this event set the universe on the road to its ultimate destiny. By considering how the universe started, and investigating the processes that occurred in the primeval phase, crucial clues can be gleaned about the far future.

The idea that the universe has not always existed is deeply ingrained in Western culture. Although the Greek philosophers considered the possibility of an eternal universe, all the major Western religions have maintained that the universe was created by God at some particular moment in the past.

The scientific case for an abrupt origin in a big bang is compelling. The most direct evidence comes from the study of the quality of light from distant galaxies. In the 1920s, the American astronomer Edwin Hubble—following up the patient observations of Vesto Slipher, an expert on nebulas who worked at the Flagstaff observatory, in

Arizona—noted that faraway galaxies appeared to be slightly redder in color than nearby ones. Hubble used the 100-inch Mount Wilson telescope to measure this reddening carefully and plotted a graph. He found that it was systematic: the farther away a galaxy lies from us, the redder it appears.

The color of light is related to its wavelength. In the spectrum of white light, blue lies at the shortwave end and red lies at the longwave end. The reddening of distant galaxies indicates that the wavelength of their light has been stretched somehow. By carefully determining the positions of characteristic lines in the spectra of many galaxies, Hubble was able to confirm this effect. He proposed that the stretching of the light waves is due to the fact that the universe is expanding. With this momentous pronouncement, Hubble laid the foundation for modern cosmology.

The nature of the expanding universe confuses many people. From the viewpoint of Earth, it seems as if the distant galaxies are rushing away from us. However, this does *not* mean that the Earth is at the center of the universe; the pattern of expansion is (on average) the same throughout the universe. Every galaxy—or, more accurately, every cluster of galaxies—moves away from every other. This is best envisaged as the stretching or swelling of the space *between* the galactic clusters rather than as the motion of the galactic clusters *through* space.

The fact that space can stretch may seem surprising, but it is a concept that has been familiar to scientists since 1915, when Einstein published his general theory of relativity. This theory proposes that gravity is actually a manifestation of the curvature, or distortion, of space (strictly, spacetime). In a sense, space is elastic, and can bend or stretch in a manner that depends on the gravitational properties of the material in it. This idea has been amply confirmed by observation.

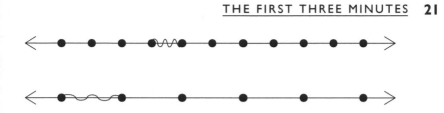

FIGURE 3.1

A one-dimensional model of an expanding universe. The buttons represent galactic clusters, and the elastic string represents space. As the string is stretched, the buttons move apart. The stretching serves to increase the wavelength of the wave propagating along the string. This corresponds to the redshift of light discovered by Hubble.

· · · · · · · · ·

The basic concept of expanding space can be understood with the help of a simple analogy. Imagine a row of buttons, representing galactic clusters, sewn onto a string of elastic (see figure 3.1). Now imagine that you are stretching the string by pulling on the ends. All the buttons move apart from one another. Whichever button you choose to consider, the neighboring buttons seem to be moving away. Nevertheless, the expansion is the same everywhere: there is no privileged center. Of course, as I have drawn it there *is* a central button, but that is irrelevant to the way the system expands. We could eliminate this detail if the string-with-buttons were infinitely long, or closed into a circle.

Viewed from any particular button, the nearest neighbors would appear to recede only half as fast as the next nearest, and so on. The farther away from your viewpoint a button is, the faster it recedes. In this type of expansion, the rate of recession is proportional to the distance—a highly significant relationship. Armed with this picture, we can now imagine light waves traveling between the buttons, or galactic clusters, in the expanding space. As space stretches, so do the waves. This explains the cosmo-logical redshift. Hubble found that the amount of redshift is proportional to the distance, just as illustrated in this simple pictorial analogy.

If the universe is expanding, it must have been more compressed in the past. Hubble's observations, and the much improved ones made since, provide a measure of the rate of expansion. If we could run the cosmic movie backward, we would find all the galaxies merging together in the remote past. From a knowledge of the pres-ent rate of expansion, we can deduce that this merged state must have occurred many billions of years ago. However, it is hard to be exact, for two reasons. First, the measurements are difficult to perform precisely and are subject to a vari-ety of errors. Even though modern telescopes have greatly increased the number of galaxies investigated, the expan-sion rate is still uncertain to within a factor of two, and is the subject of lively controversy.

Second, the rate at which the universe expands does not remain constant with time. This is due to the force of gravity, which acts between the galaxies—and, indeed, between all forms of matter and energy in the universe. Gravity acts as a brake, restraining the galaxies in their outward rush. Consequently, the expansion rate gradually diminishes with time. It follows that the universe must have been expanding faster in the past than it is now. If we plot a graph of the size of a typical region of the universe against time, we get a curve of the general form indicated by figure 3.2. From the graph, we see that the universe started out very compressed and expanding very rapidly, and the density of matter has steadily declined over time as the volume of the universe has grown. If the curve is traced all the way back to the beginning (marked 0 in the figure), it suggests that the universe originated with zero size and an infinite rate of expansion. In other words, the material that makes up all the galaxies we can see today

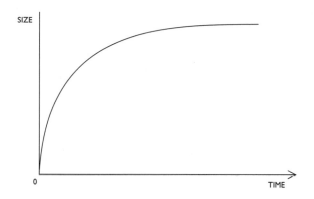

FIGURE 3.2

The rate of expansion of the universe steadily decelerates with time in approximately the manner shown. In this simple model, the rate of expansion is infinite at the point marked zero on the *time* axis. This point corresponds to the big bang.

· · · · · · · · ·

emerged from a single point, explosively fast! This is an idealized description of the so-called big bang.

But are we justified in extrapolating the curve all the way back to the beginning? Many cosmologists believe so. Given that we expect the universe to have had a beginning (for the reasons I discussed in the previous chapter), it certainly looks as though the big bang is it. If so, then the beginning of the curve marks more than merely an explosion. Remember that the expansion being graphed here is that of *space itself*, so zero volume doesn't mean merely that matter is squashed to an infinite density. It means that *space* is compressed to nothing. In other words, the big bang is the origin of space as well as of matter and energy. It is most important to realize that according to this picture there was no preexisting void in which the big bang happened.

The same basic idea applies to time. The infinite density of matter and the infinite squashing of space also mark a boundary to time. The reason is that time as well as space is stretched by gravity. Again, this effect is a consequence of Einstein's general theory of relativity and has been directly tested experimentally. The conditions at the big bang imply an *infinite* distortion of time, so that the very concept of time (and space) cannot be extended back beyond the big bang. The conclusion that seems to force itself upon us is that the big bang was the ultimate beginning of all physical things: space, time, matter, and energy. It is evidently meaningless to ask (as many people do) what happened before the big bang, or what caused the explosion to occur. There *was* no before. And where there is no time, there can be no causation in the ordinary sense.

If the big-bang theory, with its strange implications for the cosmic origin, rested only upon the evidence for the expansion of the universe, many cosmologists would probably reject it. However, important additional evidence in support of the theory came in 1965, with the discovery that the universe is bathed in heat radiation. This radiation comes at us from space with the same intensity in all directions of the sky and has been traveling more or less undisturbed since shortly after the big bang. It thus provides a snapshot of the state of the primeval universe. The spectrum of the heat radiation matches exactly the glow that exists inside a furnace that has reached a state of thermodynamic equilibrium—a form of radiation known to physicists as blackbody radiation. We are thus led to conclude that the early universe was in such a state of equilibrium, with all regions at a common temperature.

Measurements of the background heat radiation reveal it to be about three degrees above absolute zero (absolute zero is about -273°C), but the temperature changes slowly with time. As the universe expands, it cools according to a simple formula: double the radius, and the temperature falls by half. This cooling is the same effect as the red-shifting of light: both heat radiation and light consist of electromagnetic waves, and the wavelength of the heat radiation, too, stretches as the universe expands. Low-temperature radiation consists of longer waves (on average) than does high-temperature radiation. Again, running the movie backward, we see that the universe must have been very much hotter in the past. The radiation itself dates from about three hundred thousand years after the big bang, when the universe had cooled to a temperature of about 4000°C. Before this time, the primordial gas, consisting mainly of hydrogen, was an ionized plasma and therefore opaque to electromagnetic radiation. With the decline in temperature, the plasma turned into normal (un-ionized) hydrogen gas, which is transparent, allowing the radiation to propagate through it freely.

The background radiation is distinctive not just for the blackbody form of its spectrum but also for its extreme uniformity across the sky. The temperature of the radiation varies only about one part in a hundred thousand in different directions in space. This smoothness indicates that the universe must be remarkably homogeneous on a large scale, since any systematic clumping of matter into one region of space, or along one particular direction, would show up as a temperature variation. On the other hand, we know that the universe is not completely uniform. Matter is aggregated into galaxies, and the galaxies usually form clusters. These clusters are in turn arranged in superclusters. On a scale of many millions of light-years, the universe has a sort of frothy structure, with huge voids surrounded by sheets and filaments of galaxies.

The large-scale clumpiness of the universe must have grown somehow from a much smoother original state. Although various physical mechanisms might have been responsible, the most plausible explanation seems to be slow gravitational attraction. If the big-bang theory is correct, we would expect to see some evidence for the early stages of this clumping process imprinted in the cosmic background heat radiation. In 1992, a NASA satellite named COBE (for "Cosmic Background Explorer") revealed that the radiation is not precisely smooth but contains unmistakable ripples, or variations in intensity, from one place to another in the sky. These tiny irregularities seem to be the gentle beginnings of the superclustering process. The radiation has faithfully preserved the hint of the primordial agglomerations over the eons, and graphically demonstrates that the universe has not always been organized in the distinctive manner we see today. The accumulation of matter into galaxies and stars is an extended evolutionary process that began with the universe in an almost exactly uniform state.

There is a final strand of evidence that confirms the theory of a hot dense cosmic origin. Knowing the temperature of the heat radiation today, one can easily compute that at about one second after the beginning, the universe would have had a temperature of about ten billion degrees throughout. This is too hot even for composite atomic nuclei to have existed. At that time, matter must have been stripped down to its most elementary constituents, forming a soup of fundamental particles such as protons, neutrons, and electrons. However, as the soup cooled, nuclear reactions became possible. In particular, neutrons and protons were free to stick together in pairs, and these pairs in turn combined to form nuclei of the element helium. Calculations indicate that this nuclear activity lasted for about three minutes (hence the title of Steven Weinberg's book), during which time about one-quarter of the mass of material was synthesized into helium. This used up virtually all the available neutrons. The remaining uncombined protons were destined to become nuclei of hydrogen. The theory therefore predicts that the universe should consist of about 75 percent hydrogen and 25 percent helium. These figures are very much in accord with present-day measurements of the cosmic abundances of these elements.

The primordial nuclear reactions probably also produced very small amounts of deuterium, helium-3, and lithium. However, the heavier elements, which in total constitute less than 1 percent of the cosmic material, were not manufactured in the big bang. They were instead formed much later, inside stars, in a manner that I shall discuss in chapter 4.

Taken together, the expansion of the universe, the cosmic background heat radiation, and the relative abundances of the chemical elements are powerful evidence in favor of the big-bang theory. There are, nevertheless, many unanswered questions. Why, for example, is the universe expanding at precisely the rate that it is—in other words, why was the big bang so big? Why was the early universe so uniform, and the rate of expansion so similar in all directions and in all regions of space? What is the origin of the small density fluctuations found by COBE—fluctuations that are so crucial to the formation of the galaxies and galactic clusters?

In recent years, heroic efforts have been made to tackle these deeper puzzles by combining the big-bang theory with the latest ideas from high-energy particle physics. This "new cosmology," I should stress, rests on a much less secure scientific foundation than the topics I have discussed so far. In particular, the processes of interest involve particle energies vastly in excess of any that have been directly observed, and the cosmic epoch at which these processes occurred is a tiny fraction of a second after the cosmic birth. Conditions at that time were likely to have been so extreme that the only currently available guide is mathematical modeling, based almost entirely on theoretical ideas alone.

A central conjecture of the new cosmology is the possi-

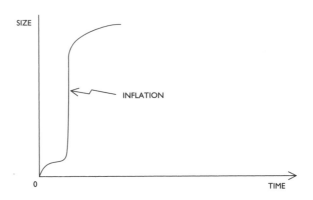

FIGURE 3.3

The inflationary scenario. In this scenario, the universe undergoes a large and sudden jump in size a very short time after originating in a bang. The vertical scale is extremely compressed. Following the inflationary phase, the expansion proceeds at a decelerating rate, in a fashion similar to that shown in figure 3.2.

· · · · · · · · ·

bility of a process called inflation. The basic idea is that at some moment during the first fraction of a second, the universe suddenly jumped in size—inflated—by a huge factor. To see what this entails, look again at figure 3.2. The curve always bends downward, indicating that while the size of any given region of space increases, it does so at a decelerating rate. By contrast, during inflation the expansion actually *accelerates*. The situation is depicted (not to scale) in figure 3.3. Initially the expansion slows, but with the onset of inflation it picks up rapidly and the curve heads skyward for a short while. Finally, the curve resumes its normal trend, but meanwhile the size of the spatial region being graphed has increased enormously (much more than shown here) compared with the equivalent position on the graph shown in figure 3.2.

Why should the universe behave in this curious way?

Remember that the downward bend of the curve is due to the attractive force of gravity acting as a brake on the expansion. An upward bend can be thought of, therefore, as a sort of antigravity, or repulsive force, causing the universe to grow in size faster and faster. Although antigravity seems an exotic possibility, some recent speculative theories suggest that such an effect might have occurred in the extreme conditions of temperature and density that prevailed in the very early universe.

Before I discuss how, let me explain why an inflationary phase helps solve some of the cosmic riddles just listed. First, the escalating expansion can give a convincing account of why the big bang was so big. The antigravity effect is an unstable, runaway process—which is to say that the size of the universe grows exponentially. Mathematically, this means that a given region of space doubles in a fixed period of time. Call this period a tick. After two ticks, the size has quadrupled; three ticks, and it has increased eightfold; ten ticks, and the region has expanded more than a thousandfold. A calculation shows that the rate of expansion at the end of the inflationary era is consistent with the observed rate of expansion today. (In chapter 6, I shall explain more precisely what I mean by this.)

The huge jump in size occasioned by inflation also provides a ready explanation for the cosmic uniformity. Any initial irregularities would be smoothed away by the stretching of space, much as the wrinkles in a balloon disappear when it is inflated. Likewise, any early variations in the expansion rate in different directions would soon be overtaken by inflation, which operates with the same vigor in all directions. Finally, the slight irregularities revealed by COBE could be attributed to the fact that inflation may not have ended at the same instant everywhere (for reasons to be discussed shortly), so some regions would have inflated slightly more than others, producing slight variations in density.

Let's put some numbers in. In the simplest version of the inflation theory, the inflationary (antigravity) force turns out to be fantastically powerful, causing the universe to double in size roughly every hundred trillion-trillion-trillionths (10^{-34}) of a second. This almost infinitesimal duration of time is what I have called a tick. After a mere hundred ticks, a region the size of an atomic nucleus would have inflated to nearly a light-year across. This is easily enough to solve the foregoing cosmological riddles.

Several possible mechanisms that might lead to inflationary behavior have been discovered by appealing to the theories of subatomic-particle physics. All these mechanisms make use of a concept known as the quantum vacuum. To understand what is involved, it is first necessary to know something about quantum physics. The quantum theory began with a discovery about the nature of electromagnetic radiation, such as heat and light. Although this radiation propagates through space in the form of waves, it can nevertheless behave as if it consisted of particles. In particular, the emission and absorption of light takes place in the form of tiny packets (or quanta) of energy, called photons. This odd amalgam of wave and particle aspects, sometimes called wave/particle duality, turned out to apply to all physical entities at the atomic and subatomic scale. Thus, entities normally thought of as particles—such as electrons, protons, and neutrons—and even whole atoms exhibit wavelike aspects under some circumstances.

A central tenet of the quantum theory is the uncertainty principle of Werner Heisenberg, according to which quantum objects do not possess sharply defined values for all their attributes. For example, an electron cannot have both a definite position and a definite momentum at the same time. Nor can it have a definite value for its energy at a definite moment of time. It is the uncertainty in energy value that concerns us here. Whereas in the macroscopic world of the engineer energy is always conserved (it cannot be created or destroyed), this law can be suspended in the subatomic quantum realm. Energy can change, spontaneously and unpredictably, from one moment to the next. The shorter the interval considered, the greater these random quantum fluctuations will be. In effect, the particle can borrow energy from nowhere, as long as the loan is paid back promptly. The precise mathematical form of Heisenberg's uncertainty principle demands that a big energy loan must be repaid very quickly, whereas smaller loans last longer.

Energy uncertainty leads to some curious effects. Among these is the possibility that a particle, such as a photon, can suddenly come into existence out of nothing, only to fade away again soon afterward. These particles live on borrowed energy, hence on borrowed time. We don't see them, because their appearance is only fleeting, but what we normally think of as empty space is really teeming with hordes of such temporarily existing particles—not just photons, but electrons, protons, and everything else. To distinguish these temporary particles from the more familiar, permanent ones, the former are called "virtual" and the latter are known as "real."

Apart from their temporary nature, virtual particles are identical to real particles. In fact, if enough energy is somehow supplied from outside the system to pay off the Heisenberg energy loan, then it is possible for a virtual particle to become a real one, and as such it is indistinguishable from any other real particle of the same species. A virtual electron, for example, typically survives for only about 10^{-21} seconds. During its brief life it does not remain at rest but may travel a distance of 10^{-11} centimeters (for comparison, an atom has a size of about 10^{-8} centimeters) before vanishing. If the virtual electron receives energy during this short time (say, from an electromagnetic field), it need not vanish after all but may continue in existence as a perfectly normal electron.

Even though we can't see them, we know that these virtual particles are "really there" in empty space because they leave a detectable trace of their activities. One effect of virtual photons, for example, is to produce a tiny shift in the energy levels of atoms. They also cause an equally tiny change in the magnetic moment of electrons. These minute but significant alterations have been very accurately measured using spectroscopic techniques.

The simple picture of the quantum vacuum given above is modified when account is taken of the fact that subatomic particles do not generally move freely but are subject to a variety of forces—the type of force depending on the type of particle involved. These forces also act between the corresponding virtual particles. It may then be that there exists more than one sort of vacuum state. The existence of many possible "quantum states" is a familiar feature of quantum physics—the best known are the various energy levels of atoms. An electron orbiting an atomic nucleus can exist in certain well-defined states with definite energies. The lowest level is called the ground state, and is stable; the higher levels are excited states, and are unstable. If an electron is knocked into a higher state, it will make one or more downward transitions back to the ground state. The excited state "decays" with a well-defined half-life.

Similar principles apply to the vacuum, which may have one or more excited states. These states would have very different energies, although they would actually look identical—that is, empty. The lowest-energy, or ground, state is sometimes called the true vacuum, reflecting the fact that it is the stable state and presumably the one that corresponds to the empty regions of the universe as observed today. An excited vacuum is referred to as a false vacuum.

False vacuums, it should be emphasized, remain a purely theoretical idea, and their properties depend a great deal on the particular theory that is being invoked. They emerge naturally, however, in most recent theories that aim to unify the four fundamental forces of nature: gravitation and electromagnetism, familiar from daily life, and two short-range nuclear forces called the weak force and the strong force. The list used to be longer: electricity and magnetism were once regarded as distinct. The process of unification began in the early nineteenth century and has advanced in recent decades. It is now known that the electromagnetic and weak nuclear forces are linked, forming a single "electroweak force." Many physicists believe that the strong force will turn out to be linked to the electroweak force, an association described in one form or another by the so-called grand unified theories. It may well be that all four forces merge into a single super-force at some deep level.

The most promising candidate for an inflationary mechanism is predicted by the various grand unified theories. A key feature of these theories is that the energy of the false-vacuum states is stupendous: typically a cubic centimeter of space would contain 10^{87} joules! Even an atomic volume in such a state would contain 10^{62} joules. Compare this with the paltry 10^{-18} joules or so that an excited atom possesses. So it would take a lot of energy to excite the true vacuum, and we would not expect to encounter a false vacuum in the universe today. On the other hand, given the extreme conditions of the big bang, these figures are plausible.

The huge energy associated with false-vacuum states has a powerful gravitational effect. This is because, as Einstein showed us, energy has mass, and it therefore exerts gravitational attraction, just as normal matter does. The enormous energy of the quantum vacuum is extremely attractive: the energy of a cubic centimeter of false vacuum would weigh 10^{67} tons, which is more than the entire observable universe today (about 10^{50} tons)! This colossal gravity is no help in producing inflation, a process that requires some sort of *anti*gravity. However, the huge false-vacuum energy is associated with an equally huge false-vacuum *pressure*, and it is this pressure that does the trick. Normally we do not think of pressure as a source of gravity, but it is. Although pressure exerts an outward mechanical force, it gives rise to an inward gravitational pull. In the case of familiar bodies, the gravitational effect of pressure is negligible compared to the effect of the body's mass. For example, less than one-billionth of your weight on Earth is due to the Earth's internal pressure. Nevertheless, the pressure effect is real, and in a system where the pressure reaches extreme values, the gravitational effect of pressure can rival that of mass.

In the case of the false vacuum, there is both a colossal energy *and* a comparably colossal pressure, so that they vie for gravitational dominance. The crucial property, however, is that the pressure is *negative*. The false vacuum doesn't push: it sucks. A negative pressure produces a negative gravitational effect—which is to say, it *antigravitates*. So the gravitational action of the false vacuum involves a competition between the huge attractive effect of its energy and the huge repulsive effect of its negative pressure. It turns out that the pressure wins, and the net effect is to create a repulsive force so large that it can blow the universe apart in a split second. It is this gargantuan inflationary push that causes the universe to double in size as rapidly as every 10^{-34} seconds.

The false vacuum is inherently unstable. Like all excited quantum states, it wants to decay back to the ground state—the true vacuum. It probably does this after a few dozen ticks. Being a quantum process, it is subject to the inevitable indeterminism and random fluctuations discussed above in connection with the Heisenberg uncertainty principle. This means that the decay will not occur uniformly throughout space: there will be fluctuations. Some theorists suggest that these fluctuations may be the source of the COBE ripples.

When the false vacuum has decayed, the universe resumes its normal decelerating expansion. The energy that had been locked up in the false vacuum is released, appearing in the form of heat. The huge distension produced by inflation had cooled the universe to a temperature very close to absolute zero; suddenly, the termination of inflation reheats it to a prodigious 10^{28} degrees. This vast reservoir of heat survives today, in grossly diminished form, as the cosmic background heat radiation. A by-product of the release of the vacuum energy is that many virtual particles in the quantum vacuum receive some of it and get promoted to real particlehood. After further processing and changes, a remnant of these primordial particles went on to provide the 10^{50} tons of matter that makes up you, me, the galaxy, and the rest of the observable universe.

If the inflationary scenario is on the right track—and many leading cosmologists believe that it is—then the basic structure and physical contents of the universe were determined by processes that were complete after a mere 10^{-32} seconds had elapsed. The postinflationary universe underwent many additional changes at the subatomic level, as the primeval material developed into the particles and atoms that constitute the cosmic stuff of our epoch, but most of the additional processing of matter was complete after only three minutes or so.

How do the first three minutes relate to the last? Just as the fate of a bullet fired toward a target depends critically on the aim of the gun, so the fate of the universe depends sensitively on its initial conditions. We shall see how the way in which the universe expanded from its primeval origins, and the nature of the material that emerged from the big bang, serve to determine its ultimate future. The beginning and the end of the universe are deeply intertwined.

CHAPTER 4

···

STARDOOM

On the night of February 23–24, 1987, a Canadian astronomer named Ian Shelton was working at the Las Campanas observatory, high in the Chilean Andes. A night assistant stepped outside briefly, and glanced idly at the dark night sky. Being familiar with the heavens, he was quick to notice something unusual. On the edge of the nebulous patch of light known as the Large Magellanic Cloud was a star. It wasn't especially bright—about the same magnitude as the those in the belt of Orion. What was significant about it was that it hadn't been there the day before.

The assistant drew Shelton's attention to the object, and within hours the news was being flashed all around the world. Shelton and his Chilean assistant had discovered a supernova. It was the first such object visible to the naked eye since Johannes Kepler recorded one in 1604. Astronomers in several countries immediately began training their instruments on the Large Magellanic Cloud. In the subsequent months, the behavior of Supernova 1987A was scrutinized in the finest detail.

Some hours before Shelton made his sensational discovery, another unusual event was being recorded in a very different place—the Kamioka zinc mine, deep beneath the ground in Japan. This was the site of a long-running experiment being conducted by physicists with an ambitious goal. Their aim was to test the ultimate stability of one of the most fundamental constituents of matter: protons. The grand unified theories developed in the 1970s predict that protons may be very slightly unstable, occasionally decaying in an exotic variant of radioactivity. If this is so, it will have profound implications for the fate of the universe, as we shall see in a later chapter.

To test for proton decay, the Japanese experimenters had filled a tank with 2000 tons of ultrapure water and stationed highly sensitive photon detectors around it. The job of the detectors was to register tell-tale flashes of light that might be attributable to the high-speed products of individual decay events. A subterranean location was chosen for the experiment in order to reduce the effects of cosmic radiation, which would otherwise swamp the detectors with spurious events.

On February 22nd, the Kamioka detectors were suddenly triggered no less than eleven times in as many seconds. Meanwhile, on the other side of the planet, a similar detector in a salt mine in Ohio was recording eight events. Since simultaneous mass suicide by nineteen protons was unthinkable, the events had to have another explanation. The physicists soon found it. Their equipment must have registered the destruction of protons by another, more conventional process: bombardment by neutrinos.

Neutrinos are subatomic particles that will play a key role in my story, so it is worth pausing to examine them in more detail. Their existence was first posited by the Austrian-born theoretical physicist Wolfgang Pauli in 1931, to explain a problematic aspect of the radioactive process known as beta decay. In a typical beta-decay event, a neutron decays into a proton plus an electron. The electron, a relatively light particle, flies off with considerable energy. The problem is that in each decay event the electron seems to have a different energy, somewhat less than the total available from the neutron's decay. Since the total energy is the same in all cases, it seems as if the final energy differs from the original. This won't do, as it is a basic law of physics that energy is conserved, so Pauli suggested that the missing energy was being conveyed away by an invisible particle. Early attempts to detect these particles failed, and it became clear that if they existed they must have incredible penetrating power. As any sort of electrically charged particle would readily be trapped by matter, Pauli's particle had to be electrically neutral—hence the name "neutrino."

Although at the time nobody had actually spotted a neutrino, theorists were able to figure out more of their properties. One of these properties concerns the neutrino's mass.

The concept of mass is a subtle one when it comes to fast-moving particles. That is because the mass of a body is not a fixed quantity but depends on the body's speed. For example, a 1-kilogram lead ball would weigh 2 kilograms if it moved at 260,000 kilometers a second. The key factor here is the speed of light. The closer an object's speed gets to the speed of light, the more massive it becomes, and this rise in mass is without limit. Because mass is variable in this way, when physicists talk about the mass of a subatomic particle they refer to its mass *at rest*, to avoid confusion. If the particle is moving at close to the speed of light, its *actual* mass may be many times its rest mass: inside large particle accelerators, circulating electrons and protons may have many thousands of times their rest masses.

A clue to the value of the rest mass of the neutrino comes from the fact that a beta-decay event will sometimes eject an electron with nearly all the available energy, leaving almost none for the neutrino. This means that neutrinos can exist with essentially zero energy. Now, according to Einstein's famous formula $E = mc^2$, energy E and mass m are equivalent, so zero energy implies zero mass. This means that the neutrino is likely to have a very small, possibly zero, rest mass. If the rest mass is truly zero, then the neutrino will travel at the speed of light. In any case, it is likely to be found traveling at very close to the speed of light.

A further property concerns the way that subatomic particles spin. Neutrons, protons, and electrons are always found to be spinning. The magnitude of this spin is a certain fixed quantity, and in fact it is the same quantity for all three. Spin is a form of angular momentum, and there is a law of conservation of angular momentum—a law as basic as the law of conservation of energy. When a neutron decays, its spin must be preserved in the decay products. If the electron and the proton were spinning in the same direction, their spins would add to make twice that of the neutron. On the other hand, if they were counterrotating, the spins would cancel to give zero in total. Either way, the total spin of an electron and a proton alone could not be equal to that of the neutron. However, when account is taken of the existence of a neutrino, the books can be made to balance nicely by assuming that the neutrino possesses the same spin as the other particles. Then two of the three decay products can spin in the same direction, while the third counterrotates.

So without ever having detected a neutrino, physicists were able to deduce that it must be a particle with zero electric charge, identical spin to the electron, little or no rest mass, and such a feeble interaction with ordinary matter as to leave almost no trace of its passage. In short, it is a sort of spinning ghost. Not surprisingly, it took about twenty years after Pauli conjectured the existence of neutrinos for them to be definitively identified in the laboratory. They are created in such copious quantities in nuclear reactors that despite their extraordinary elusiveness it is possible to detect the occasional representative.

The arrival of a burst of neutrinos in the Kamioka mine at the same time as the appearance of Supernova 1987A was undoubtedly not simply a coincidence, and the concurrence of the two events was seized upon by scientists as crucial confirmation of the theory of supernovas: a burst of neutrinos was exactly what astronomers had long expected from a supernova.

Although the word "nova" means "new" in Latin, Supernova 1987A was not the birth of a new star. It was, in fact, the death of an old one in a spectacular explosion. The Large Magellanic Cloud, in which the supernova appeared, is a minigalaxy located about a hundred and seventy thousand light-years away. This is close enough to the Milky Way to make it a sort of satellite of our galaxy. It is visible to the unaided eye as a fuzzy patch of light in the Southern Hemisphere, but large telescopes are needed to reveal its individual stars. Only hours after Shelton's discovery, Australian astronomers were able to identify which star among the few billion contained in the Large Magellanic Cloud was the one that blew up; they accomplished this feat by inspecting previous photographic plates of that region of the sky. The stricken star was of a type known as a B3 blue supergiant, and its diameter was about forty times that of the sun. It even had a name: Sanduleak -69 202.

The theory that stars can explode was first investigated by the astrophysicists Fred Hoyle, William Fowler, and Geoffrey and Margaret Burbidge, in the mid-1950s. To understand how a star arrives at such a cataclysm it is necessary to know something about its internal workings. The most familiar star is the sun. In common with most stars, the sun seems changeless; however, this belies the fact that it is locked in a ceaseless struggle with the forces of destruction. All stars are balls of gas held together by gravity. If gravity were the only force at work, they would instantly implode under their own immense weight and vanish within hours. The reason they don't is that the inward force of gravity is balanced by the outward force of the pressure of the compressed gas in the stellar interior.

There is a simple relation between the pressure of a gas and its temperature. When a gas of fixed volume is heated, the pressure normally rises in proportion to the temperature. Conversely, when the temperature falls, so does the pressure. The interior of a star has an enormous pressure because it is so hot—many millions of degrees. The heat is produced by nuclear reactions. For most of its lifetime, the principal reaction that powers a star is the conversion of hydrogen into helium by fusion. This reaction requires a very high temperature to overcome the electric repulsion that acts between nuclei. Fusion energy can sustain a star for billions of years, but sooner or later the fuel runs low, and the reactor starts to falter. When this happens, the pressure support is threatened and the star begins to lose its long battle with gravity. A star essentially lives on borrowed time, staving off gravitational collapse by marshaling its reserves of fuel. But every kilowatt that flows away from the stellar surface into the depths of space serves to hasten the end.

It is reckoned that the sun can burn for about ten billion years on the hydrogen it started out with. Today, at about five billion years of age, our local star has burned up nearly half its reserves. (No need to panic just yet.) The rate at which a star consumes nuclear fuel depends sensitively on its mass. Heavier stars burn fuel much faster— they must, because they are bigger and brighter, and so radiate more energy. The extra weight squeezes the gas to a higher density and temperature, increasing the fusion reaction rate. A star with ten solar masses, for example, will burn up most of its hydrogen within as little as ten million years.

Let us follow the fate of such a massive star. Most stars

start out composed mainly of hydrogen. Hydrogen "burning" consists of the fusion of hydrogen nuclei—the hydrogen nucleus is a single proton—to form nuclei of the element helium, each consisting of two protons and two neutrons. (The details are complicated and need not concern us here.) Hydrogen "burning" is the most efficient source of nuclear energy, but it is not the only one. If the core temperature is high enough, helium nuclei can fuse to form carbon, and further fusion reactions lead to oxygen, neon, and other elements. A massive star can generate the necessary internal temperatures—amounting to over a billion degrees—for this chain of successive nuclear reactions to proceed, but the returns steadily diminish. With each new element forged, the energy released declines. The fuel is burned up faster and faster, until the composition of the star changes monthly, then daily, then hourly. Its interior resembles an onion, with the layers being the successive chemical elements synthesized at an ever more frenetic pace. Externally, the star balloons to an enormous size, larger than that of our entire solar system, becoming what astronomers call a red supergiant.

The end of the nuclear-burning chain is marked by the element iron, which has a particularly stable nuclear structure. The synthesis of elements heavier than iron by nuclear fusion actually costs energy rather than liberates it, so that by the time a star has synthesized a core of iron, it is doomed. Once the central regions of the star can no longer produce heat energy, the odds tip fatally in favor of the force of gravity. The star teeters on the edge of catastrophic instability, eventually toppling into its own gravitational pit.

What happens, and happens fast, is this. The iron core of the star, no longer capable of producing heat by nuclear burning, cannot support its own weight, and it contracts so forcefully under gravity that the very atoms are crushed. Eventually the core reaches nuclear densities, at which a thimble will accommodate nearly a trillion tons of matter. At this stage, the core of the stricken star will typically be two hundred kilometers across, and the solidity of the nuclear material will cause it to bounce. So strong is the gravitational pull that this titanic rebound takes but a few milliseconds. As the drama unfolds in the center of the star, the surrounding layers of stellar material collapse onto the core in a sudden, calamitous convulsion. Traveling inward at tens of thousands of kilometers per second, the trillions upon trillions of tons of imploding material encounter the rebounding highly compact core, harder than a diamond wall. What follows is a collision of staggering violence, sending a huge shock wave outward through the star.

Accompanying the shock wave is a tremendous pulse of neutrinos, liberated suddenly from the inner regions of the star during its final nuclear transmutation—a transmutation in which the electrons and protons of the star's atoms are crushed together to form neutrons. The core of the star effectively becomes a giant ball of neutrons. Together, the shock wave and the neutrinos transport a vast quantity of energy outward through the overlying layers of the star. Absorbing much of this energy, the outer layers of the star explode in a nuclear holocaust of unimaginable fury. For a few days, the star shines with the intensity of ten billion suns, only to fade away a few weeks later.

Supernovas occur, on average, two or three times a century in a typical galaxy like the Milky Way, and they have been recorded in history by astonished astronomers. One of the most famous was noted by Chinese and Arab observers in A.D. 1054 in the constellation of Cancer, the crab. Today, the shattered star appears as a ragged cloud of expanding gas known as the Crab Nebula.

The explosion of Supernova 1987A illuminated the universe with an invisible flash of neutrinos. It was a pulse of staggering intensity. Every square centimeter of Earth—even though it is a hundred and seventy thousand light-years away from the explosion—was pierced by a hundred billion neutrinos, its inhabitants blissfully unaware that they had been momentarily penetrated by many trillions of particles from another galaxy. But the Kamioka and Ohio proton-decay detectors stopped nineteen of them. Without this equipment, the neutrinos would have passed by unnoticed, as they did in 1054.

Although a supernova spells death to the star concerned, the explosion has a creative aspect to it. The enormous release of energy heats the outer layers of the star so effectively that for a brief time further nuclear-fusion reactions are possible—those reactions that soak up rather than release energy. Heavy elements beyond iron—such as gold, lead, and uranium—are forged in that final and most intense stellar furnace. These elements, along with the lighter ones, such as carbon and oxygen, that were created in the earlier stages of nucleosynthesis, are blasted into space, there to mingle with the detritus of countless other supernovas. Over the ensuing eons, these heavy elements are scooped up into new generations of stars and planets. Without the manufacture and dissemination of these elements, there could be no planets like the Earth. Life giving carbon and oxygen, the gold in our banks, the lead sheeting on our roofs, the uranium fuel rods of our nuclear reactors—all owe their terrestrial presence to the death throes of stars that vanished well before our sun existed. It is an arresting thought that the very stuff of our bodies is composed of the nuclear ash of long-dead stars.

A supernova explosion does not completely destroy the star. Although most of the material is dispersed by the cataclysm, the imploded core that triggered the event remains in place. Its fate, however, is also a touch-and-go affair. If the mass of the core is fairly low—say, one solar mass—then it will form a ball of neutrons the size of a small city. Most likely, this "neutron star" will be spinning frenetically, perhaps over 1000 revolutions per second, or 10 percent of the speed of light at the surface. It acquires this dizzy spin because the implosion hugely amplifies the relatively slow rotation of the original star; this is the same principle that causes ice skaters to spin faster when they retract their arms. Astronomers have detected many such rapidly rotating neutron stars. But the rotation rate gradually slows as the object loses energy. The neutron star in the middle of the Crab Nebula, for example, has now slowed to 33 revolutions per second.

If the mass of the core is somewhat larger—say, several solar masses—it cannot settle down as a neutron star. The force of gravity is so strong that even neutronic matter—the stiffest-known substance—cannot resist further compression. The stage is then set for an event more awesome and more catastrophic than the supernova. The core of the star continues to collapse, and in less than a millisecond it creates a black hole and disappears into it.

The fate of a massive star, then, is to blow itself to bits, leaving as a remnant either a neutron star or a black hole surrounded by diffuse ejected gases. Nobody knows how many stars have already succumbed in this manner, but the Milky Way alone could contain billions of these stellar corpses.

As a child, I had a morbid fear that the sun might explode. There is no danger, however, of its becoming a supernova. It is too small. The fate of light stars is altogether much less violent than that of their massive cousins. In the first place, the nuclear processes that devour fuel proceed at a more sedate pace; indeed, a dwarf star at the bottom end of the stellar-mass range may shine steadily for a trillion years. In the second place, a light star cannot generate internal temperatures high enough to synthesize iron, and hence to unleash a catastrophic implosion.

The sun is a typical fairly low-mass star, steadily burning through its hydrogen fuel and turning its interior into helium. The helium mostly resides in a central core that is inert as far as nuclear reactions are concerned; the fusion takes place at the surface of the core. Therefore the core itself is unable to contribute to the crucial heat generation needed to hold the sun up in the face of crushing gravitational forces. To prevent collapse, the sun must expand its nuclear activity outward, in search of fresh hydrogen. Meanwhile, the helium core gradually shrinks. As the eons slip by, the sun's appearance will imperceptibly alter as a result of these internal changes. It will swell in size, but its surface will cool somewhat, giving it a reddish hue. This trend will continue until the sun turns into a red giant star, perhaps five hundred times as big as it is now. Red giants are familiar to astronomers, and several well-known bright stars in the night sky, such as Aldebaran, Betelgeuse, and Arcturus, fall into this category. The red-giant phase marks the beginning of the end for a low-mass star.

Although a red giant is relatively cool, its large size gives it a huge radiating surface, which means a greater overall luminosity. The sun's planets will face a hard time, some four billion years down the track, as the increased heat flux assails them. The Earth will become uninhabitable long before this, its oceans boiled away and its atmosphere stripped. As the sun grows ever more distended, it will engulf Mercury, then Venus, and finally Earth within its fiery envelope. Our planet will be reduced to a cinder, doggedly clinging to its orbit even after incineration; the density of the sun's red-hot gases will be so low that conditions will approximate a vacuum, exerting little drag on Earth's motion.

Our very existence in the universe is a consequence of the extraordinary stability of stars like the sun, which can burn steadily with little change for billions of years, long enough to allow life to evolve and flourish. But in the red-giant phase this stability will come to an end. The succeeding stages in the career of a star like the sun are complicated, erratic, and violent, with relatively sudden changes of behavior and appearance. Aging stars may spend millions of years pulsating, or sloughing off shells of gas. The helium in the star's core may ignite to form carbon, nitrogen, and oxygen—thereby providing vital energy that will sustain the star a while longer. By blowing off its outer envelope into space, a star can end up stripped down to its carbon-oxygen core.

Following this period of complicated activity, low- and medium-mass stars inevitably succumb to gravity and shrink. The shrinkage is remorseless, and continues until the star is compressed to the size of a small planet, becoming an object known to astronomers as a white dwarf. Because white dwarfs are so small, they are extremely dim, in spite of the fact that their surface temperatures can be much greater than that of the sun. None are visible from Earth without the aid of a telescope.

It is our sun's destiny to become a white dwarf in the far future. When the sun reaches that phase, it will continue to remain hot for many billions of years; its huge bulk will be so compacted that it will trap its internal heat more efficiently than the best-known insulator. However, because the internal nuclear furnace will have shut down for good, there will be no reserves of fuel to replenish the slow leakage of heat radiation into the cool depths of space. Very, very slowly, the dwarf remnant of what was once our mighty sun will cool and dim, until it embarks on its final metamorphosis, gradually solidifying into a crystal of extraordinary rigidity. Eventually it will fade out completely, merging quietly into the blackness of space.

CHAPTER 5

NIGHTFALL

The Milky Way blazes with the light of a hundred billion stars, and every one of them is doomed. In ten billion years, most that we see now will have faded from sight, snuffed out from lack of fuel, victims of the second law of thermodynamics.

But the Milky Way will still glow with starlight, for even as stars die, new stars are born to take their place. In the galaxy's spiral arms, such as the one our sun is located in, gas clouds become compressed, collapse under gravity, fragment, and produce a cascade of stellar births. A glance at the constellation of Orion reveals the activity of such a stellar nursery. The fuzzy blob of light in the center of Orion's sword is not a star but a nebula—a huge cloud of gas studded with bright young stars. By observing infrared radiation rather than visible light, astronomers looking at this nebula have recently glimpsed stars in the very first stages of formation, still surrounded by obscuring gas and dust.

The formation of stars will continue in the spiral arms of our galaxy as long as there is enough gas. The gas content of the galaxy is partly primordial—material that has not yet aggregated into stars—and partly gas that has been ejected from stars in supernovas, stellar winds, small explosive outbursts, and other processes. Obviously, the recycling of matter cannot go on indefinitely. As the old stars die and collapse to become white dwarfs, neutron stars, or black holes, they will be unable to replenish the interstellar gases. Slowly the primordial material will become incorporated into stars, until it, too, is totally depleted. As these latter-day stars pass through their life cycles and die, the galaxy will grow inexorably dimmer. The fade-out will be protracted. Many billions of years will elapse before the smallest, youngest stars complete their nuclear burning and shrink into white dwarfs. But with slow, agonizing finality perpetual night will surely fall.

A similar fate awaits all the other galaxies scattered across the ever-widening chasms of space. The universe, currently aglow with the prolific energy of nuclear power, will eventually exhaust this valuable resource. The era of light will be over forever.

The end of the universe will not come when the cosmic lights go out, however, for there is another source of energy even more powerful than nuclear reactions. Gravity, the weakest of nature's forces at the atomic level, becomes dominant on the astronomical scale. It may be relatively gentle in its effects, yet it is utterly persistent. For billions of years, stars shore themselves up against their own weight by nuclear burning. But all the while gravity is waiting to claim them.

The gravitational force between two protons in an atomic nucleus is a mere ten-trillion-trillion-trillionth (10^{-37}) of the strong nuclear force. But gravity is cumulative. Every additional proton in a star adds to the total weight. Eventually, the gravitational force is overwhelming. And this overwhelming force is the key that unlocks immense power.

No object illustrates the power of gravitation more graphically than a black hole. Here, gravitation has triumphed totally, crushing a star to nothing and leaving an imprint in the surrounding spacetime in the form of an infinite time warp. There is a fascinating thought experiment concerning black holes. Imagine dropping a small object—for example, a 100-gram weight—into a black hole from a great distance. The weight will plunge out of sight into the hole and become irretrievably lost. It leaves a vestige of its erstwhile existence, however, in the structure of the hole, which becomes very slightly larger as a result of swallowing the weight. A calculation shows that if the ball is dropped from a great distance into the hole, then the hole will gain an amount of mass equal to the original mass of the weight. No energy or mass escapes.

Now consider a different experiment, in which the weight is lowered slowly toward the hole. This could be accomplished by fixing a string to it, passing the string over a pulley onto a drum, and allowing the string to unwind. (See figure 5.1. I am assuming that the string doesn't stretch or weigh anything, which is a fiction, but this is to avoid complicating the discussion.) As the weight is lowered, it can deliver energy—for example, by turning an electric generator attached to the drum. The closer the weight gets to the surface of the black hole, the stronger will be the hole's gravitational pull on the weight. As the downward force rises, the weight does more and more work on the generator. A simple calculation reveals how much energy the weight will have delivered to the generator by the time it reaches the surface of the black hole. In the ideal case, the answer turns out to be the *entire* rest-mass energy of the weight. (I have explained the concept of rest mass on page 31.)

Recall Einstein's famous $E = mc^2$ formula, which tells us that a mass m possesses an amount of energy mc^2. Using a black hole, one could in principle recover the lot. In the case of a 100-gram weight, the lot means about three billion kilowatt-hours of electricity. By way of compari-

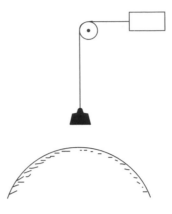

FIGURE 5.1

In this idealized thought experiment, a weight is slowly lowered on a string toward the surface of a black hole, using a fixed pulley system (fixture not shown). As a result, the descending weight performs work and delivers energy to the box. The total energy delivered approaches the entire rest-mass energy of the weight, as the weight approaches the surface of the black hole.

· · · · · · · · ·

son, when the sun burns 100 grams of fuel by nuclear fusion, it delivers less than 1 percent of this figure. So, in principle, gravitational-energy release could be over a hundred times more powerful than the thermonuclear fusion that powers stars.

Of course, the two contrived situations described here are totally unrealistic. Undoubtedly objects are continually falling into black holes, but never dangling from pulleys in the most efficient manner for energy extraction. In practice, some value intermediate between zero and 100 percent of the rest-mass energy is emitted. The actual fraction depends on the physical circumstances. Over the past couple of decades, astrophysicists have studied a wide range of computer simulations and other mathematical models in an attempt to understand the behavior of gas as it swirls into a black hole and to estimate the quantity and pattern of energy released. The physical processes involved are very complicated; nevertheless, it is clear that enormous amounts of gravitational energy can pour out of such systems.

A single observation is worth a thousand calculations, and astronomers have made extensive searches for objects that might be black holes in the process of swallowing matter. Although a completely convincing candidate black hole has not yet been found, one system that looks very promising is located in the constellation of Cygnus and is known as Cygnus X-1. An optical telescope reveals a large, hot star of the sort known as a blue giant, on account of its color. Spectroscopic studies indicate that the blue star is not alone; it executes a rhythmic wiggle, an indication that it is being periodically attracted by the gravity of a nearby object. Evidently the star and another body are in close orbit around each other. Optical telescopes, however, reveal no sign of the companion: it is either a black object or a very dim compact star. This is suggestive of a black hole, but by no means proof.

A further clue comes from estimates of the mass of the dark body. This can be deduced from Newton's laws, once we know the mass of the blue giant star—which we can estimate because of the close relationship between the mass of a star and its color: blue stars are hot and therefore have high mass. Calculations indicate that the unseen companion object has a mass equivalent to several suns. It is clearly not a normal small and dim star, so it must be a collapsed massive star—either a white dwarf, a neutron star, or a black hole. But there are basic physical reasons why a compact object this massive cannot be a white dwarf or a neutron star. The problem concerns the intense gravitational field that tries to crush the object. Total collapse to a black hole can be avoided only if some sort of internal pressure exists, strong enough to counter the squeezing force of gravity. But if the collapsed object is several solar masses, no known force can resist the crushing weight of its material. Indeed, if the core of the star were stiff enough to avoid being crushed, then the speed of sound in the material would have to exceed the speed of light. Since this is contrary to the theory of special relativity, most physicists and astronomers believe that the formation of a black hole is inevitable under these circumstances.

The clinching piece of evidence that Cygnus X-1 contains a black hole comes from another observation entirely. The designation X-1 was given because the system is a strong source of X rays, which can be detected by sensors carried aboard satellites. Theoretical models give a convincing account of these X rays based on the assumption that the dark companion object in Cygnus X-1 is a black hole. The computed gravitational field of the hole is strong enough to suck material off the blue giant star. As the abducted gases are drawn toward the hole—and eventual oblivion—the orbital rotation of the system would cause the infalling material to swirl around the black hole and form a disk. A disk of this sort cannot be completely stable, because the material near the center orbits the black hole much faster than the material near the rim, and viscous forces will try to smooth out this differential rotation. As a result, the gas heats up to a temperature high enough to emit not merely light but X rays. The loss of orbital energy this represents causes the gas to spiral slowly into the hole.

The evidence for a black hole in Cygnus X-1 therefore relies on a fairly long chain of reasoning, involving both observational details and theoretical modeling. This is typical of the nature of much astronomical research these days; no single piece of evidence is compelling, but the various studies of Cygnus X-1 and a number of similar systems, taken together, strongly suggest the presence of a black hole. Certainly the black-hole explanation is the neatest and least contrived.

Effects even more spectacular can be expected from the activities of bigger black holes. It now seems likely that many galaxies contain supermassive black holes in their centers. Evidence for this is the rapid movement exhibited by stars in these galactic cores; the stars are apparently being drawn toward a strongly attractive, highly compact object. Estimates of the mass of such possible objects range from ten million to a billion solar masses; this would give them a voracious appetite for any stray matter in their vicinity. Stars, planets, gas, and dust are probably all falling prey to these monsters. The violence of the infall process should in some cases be great enough to disturb the entire structure of the galaxy. Astronomers are familiar with many varieties of active galactic nuclei. Some galaxies give the appearance of literally exploding; many are powerful sources of radio waves, X rays, and other forms of energy. Most distinctive are a class of active galaxies that sprout enormous jets of gas—jets thousands or even millions of light-years in length. The energy output of some of these objects is amazing. For example, very distant quasars—the name is short for "quasi-stellar objects"—may emit as much energy as thousands of galaxies, yet from a region as small as one light-year across, giving them the superficial appearance of stars.

Many astronomers believe that the central engines of all these seriously disrupted objects are huge rotating black holes, which are in the process of ingesting material in their vicinity. Any stars that approach a black hole are likely to be torn apart by the hole's gravity or to collide with other stars and break up. As in the case of Cygnus X-1, but on a much larger scale, the distributed material would probably form a disk of hot gas that orbits the hole and slowly sinks inward. In May 1994 it was reported that the Hubble space telescope had revealed a rapidly rotating disk of gas at the center of galaxy M87. The observations strongly suggest the presence of a supermassive black hole.

It may happen that the copious energy released from a disk of gas flowing onto a black hole is channeled along the spin axis of the hole, producing a pair of oppositely directed jets, as often observed. The mechanism of this energy release, and the formation of jets, is likely to be very complicated, involving electromagnetic, viscous, and other forces as well as gravity. The topic remains the subject of intensive theoretical and observational work.

What about the Milky Way? Is it possible that our own galaxy will be disrupted in this manner? The center of the Milky Way lies thirty thousand light-years away, in the constellation of Sagittarius. The inner regions are obscured by large clouds of gas and dust, but radio, X-ray, gamma-ray, and infrared instruments have enabled astronomers to discern the existence of a highly compact, energetic object called Sagittarius A*. No larger than a few billion kilometers across (small by astronomical standards), Sagittarius A* is nevertheless the most powerful radio source in the galaxy. Its position coincides with that of a very intense infrared source, and is also close to an unusual X-ray object. Although the situation is complicated, it seems increasingly likely that at least one massive black hole lurks there and is responsible for some of the observed phenomena. However, the mass of the hole is probably ten million solar masses at most, putting it well at the bottom of the supermassive range. There is no evidence for the sort of violent emission of energy and material that is occurring in some other galactic nuclei, but this may be because the black hole is currently in a quiescent phase. It could flare up at some stage in the future—perhaps if it receives a greater supply of gas—though it would probably not be as disruptive as many other known systems. What effect such a flare-up would have on the stars and planets in the spiral arms of the galaxy is unclear.

A black hole will continue to release the rest-mass energy of sacrificed matter as long as there is material in the vicinity of the hole to feed it. Over time, more and more matter will get swallowed up by black holes, and the holes will grow larger and hungrier as a result. Even stars in very distant orbits around a massive hole will eventually succumb. The reason is an extremely weak yet ultimately decisive phenomenon known as gravitational radiation.

Shortly after he formulated his general theory of relativity in 1915, Einstein discovered a remarkable property of the gravitational field. From a study of the field equations of the theory, he found that they predicted the existence of wavelike gravitational oscillations that propagate at the speed of light through empty space. This gravitational radiation is reminiscent of electromagnetic radiation, such as light and radio waves. However, although it can carry a lot of energy, gravitational radiation differs from electromagnetic radiation in the strength with which it disturbs matter. Whereas a radio wave is readily absorbed by a structure as delicate as a wire mesh, a gravitational wave interacts so weakly that it will pass right through the Earth with scarcely any diminution. If you could make a gravitational laser, you would need a trillion-kilowatt beam to boil a kettle of water as efficiently as a kilowatt electric coil. The comparative feebleness of gravitational radiation can be traced to the fact that gravitation is by far the weakest of the known forces of nature. The ratio of gravitational to electric forces in an atom, for example, is about 10^{-40}. The only reason we notice gravity at all is because its effects are cumulative, so it predominates in large objects such as planets.

Not only are gravitational waves exceedingly feeble in their effects but their production is also a muted affair. In principle, gravitational radiation is produced whenever masses are disturbed. For example, the motion of the Earth around the sun emits a continuous train of gravitational waves, but the total power output is a mere milliwatt! This energy drain causes the Earth's orbit to decay, but at a ludicrously slow rate: about a thousand-trillionth of a centimeter per decade.

The situation is dramatically different, though, for massive astronomical bodies moving close to the speed of light. Two sorts of phenomena are likely to lead to important gravitational-radiation effects. One is a sudden, violent event—a supernova, say, or the collapse of a star to form a black hole. Such an event results in the emission of a short-lived pulse of gravitational radiation, lasting perhaps a few microseconds and typically carrying away 10^{44} joules of energy. (Compare this with the sun's output of heat, which is about 3×10^{26} joules per second.) The other phenomenon is the high-speed motion of massive objects in orbit about each other. For example, a closely spaced binary star system will generate a large continuous flux of gravitational radiation. This process is especially efficient if the orbiting stars are collapsed objects, such as neutron stars or black holes. In the constellation of Aquila, there are two neutron stars orbiting only a few million kilometers from each other. Their gravitational fields are so strong that each orbit is completed in under eight hours, so the stars are moving at an appreciable fraction of the speed of light. This unusually rapid motion greatly amplifies the rate of gravitational-wave emission, and causes the orbit to decay by a measurable amount each year (about 75 microseconds alteration in the period). The emission rate will escalate as the stars spiral together. They are destined to plow into each other three hundred million years from now.

Astronomers estimate that a binary system of this sort coalesces roughly once every hundred thousand years per galaxy. So compact are the objects, and so intense are their gravitational fields, that during the last moments before the stars impact they will orbit each other thousands of times per second, and the frequency of the gravitational wave will shoot up in a characteristic chirp. Einstein's formulas predict that the gravitational power output will be prodigious in this final phase, and the orbit will rapidly collapse. The shape of the stars will be severely distorted by mutual gravitational pull, so that by the time they touch they will look like giant whirling cigars. The resulting coalescence will be a messy affair, the two stars merging to form a complicated, madly cavorting mass, which will also emit gravitational radiation prolifically until it settles into a roughly spherical form, ringing and wobbling like a monstrous bell in a distinctive pattern of vibration. These oscillations, too, will produce a certain amount of gravitational radiation, draining the object of still more energy, until it quiets down and eventually becomes inert.

Although the rate of energy loss is relatively slow, the emission of gravitational radiation is likely to have profound long-term effects on the structure of the universe. It is therefore important that scientists try to confirm their ideas about gravitational radiation by observation. The studies of the binary neutron-star system in Aquila show that the orbit is decaying at precisely the rate predicted by Einstein's theory. This system therefore provides direct evidence for the emission of gravitational radiation. A more decisive test, however, requires the detection of such radiation in a laboratory on Earth. Many research groups have built equipment designed to register the fleeting passage of a burst of gravitational waves, but to date none of these devices has been sensitive enough to detect any, and it is likely that we must await a new generation of detectors before the existence of gravitational radiation can be completely confirmed.

The coalescence of two neutron stars may produce either a larger neutron star or a black hole. The coalescence of a neutron star and a black hole, or of two black holes, must produce a single black hole. This process would be accompanied by a loss of gravitational-wave energy similar to that in the case of binary neutron stars, followed by complicated ringing and wobbling motions, which would be slowly damped away by the gravitational-wave power loss.

It is interesting to explore the theoretical limits to the gravitational energy that could be extracted from two black holes during coalescence. The theory for these processes was worked out by Roger Penrose, Stephen Hawking, Brandon Carter, Remo Ruffini, Larry Smarr, and others in the early 1970s. If the holes are nonrotating and identical in mass, about 29 percent of the total rest-mass energy can be liberated. This need not be entirely in the form of gravitational radiation if the black holes were in some way manipulated—by some sort of advanced technology, say—but in a natural merger most of the energy released would be in this highly inconspicuous form. If the holes were spinning at the maximum rate allowed by the laws of physics (roughly, at the speed of light) and merged, counterrotating, along their spin axes, then 50 percent of the mass energy could be emitted.

Even this sizable fraction is not the theoretical maximum. It is possible for a black hole to carry electric charge. A charged black hole has an electric field as well as a gravitational field, and both can store energy. If a black hole with a positive charge encounters one with a negative charge, a "discharge" occurs, releasing electromagnetic as well as gravitational energy in the process.

There is a limit to this discharge, since a black hole of a given mass can carry an electric charge up to some maximum value only. For a nonrotating hole, this value is set by the following consideration. Imagine two identical holes having the same charge. The gravitational fields of the holes will cause a force of attraction between them, while the electric fields will cause a force of repulsion (like charges repel). When the charge-to-mass ratio reaches a critical value, these two opposing forces will exactly balance, and there will be no net force between the black holes. It is this condition that marks the limit to the amount of electric charge that a black hole can contain. You might wonder what would happen if you tried to increase the charge on a black hole above this maximum value. One way to attempt this would be to force more charge down the hole. This procedure will serve to increase the electric charge, but the work done in overcoming the electric repulsion uses energy, and this energy is delivered to the hole. Because energy has mass (remember $E = mc^2$) the hole gets more massive, and hence bigger. A simple calculation shows that the mass goes up by more than the charge in this process, so the charge-to-mass ratio actually decreases, and the attempt to beat the limit fails.

The electric field of a charged black hole contributes to the total mass of the hole. In the case of a hole carrying the maximum allowed charge, the electric field represents half the mass. If two nonrotating holes carry the maximum charge but of *opposite* sign, they will attract each other both gravitationally and electromagnetically. When they merge, the electric charges will neutralize, and the electrical energy can be extracted. Theoretically, it can be as much as 50 percent of the total mass energy in the system.

The absolute upper limit to energy extraction will obtain if both the holes are rotating and if they carry opposite electric charges, each to the maximum value. Then as much as two-thirds of the total mass energy can be released. Of course, such values are of theoretical interest only, because in practice a black hole is unlikely to carry a large electric charge, nor are two holes likely to merge in the optimal manner, unless they are made to do so by an advanced technological community. However, even the inefficient coalescence of two black holes will probably produce an almost instant energy release, amounting to a significant fraction of the total mass energy of the objects concerned. This can be compared to the paltry 1 percent of mass energy that stars emit by nuclear fusion over their multibillion-year lives.

The significance of these gravitational processes is that, far from dying, a burnt-out star has the potential to release vastly more energy as a collapsed cinder than from thermonuclear processes as a glowing ball of gas. When this fact was recognized about twenty years ago, the physicist John Wheeler—the man who originally coined the term "black hole"—conceived of a hypothetical civilization whose escalating energy requirements led it to abandon its star and take up residence around a rotating black hole. Every day, the waste products of the community are loaded into trucks and dispatched toward the hole on a carefully computed trajectory. Close to the hole, the contents of the trucks are released, tipping the waste into the hole, by which means it is disposed of for good. The infalling material, traveling along a rotating path counter to the spin of the hole, has the effect of braking the spin slightly. The rotational energy of the hole is thereby released, and can be harnessed by the civilization to power its industry. The process therefore has the double virtue of completely eliminating all waste products by turning them into pure energy! In this way, the civilization can release on demand a much greater supply of energy from the dead star than that star ever emitted in its luminous phase.

Although the harnessing of a black hole's power is a science-fiction scenario, a lot of matter will end up inside black holes naturally—either as part of the star that collapses to form the hole or as debris swallowed during a chance encounter. Whenever I give lectures on black holes, people always want to know what happens to something that enters one. The short answer is, We don't know. Our understanding, such as it is, of black holes is based almost entirely on theoretical considerations and mathematical modeling. Indeed, by definition we cannot observe the interior of a black hole from the outside, so even if we had good observational access to a black hole (which we don't), we could never know what was going on inside it. Nevertheless, the theory of relativity, which predicts the existence of black holes in the first place, can also be used to predict what would happen to an astronaut who falls into one. What follows is a summary of those theoretical deductions.

The surface of the hole is really only a mathematical construct—there is no actual membrane there, only empty space. The infalling astronaut would notice nothing especially different as she or he crosses into the hole. However, the surface does have a certain—and rather dramatic—physical significance. Inside the hole, gravity is so strong that it traps light, pulling outgoing photons back in again. This means that light cannot escape from the hole, which is why it appears black from the outside. Because no physical object or information can travel faster than light, nothing can escape from the black hole once this border has been crossed. Events that occur within the hole are forever hidden from external observers. For this reason, the surface of the hole is referred to as an "event horizon"—because it separates events on the outside, which can be witnessed from afar, from those on the inside, which cannot. The effect, however, is only one-way. The astronaut within the event horizon can still see the universe outside, even though nobody out there can see the astronaut.

As the astronaut plunges deeper into the hole, the gravitational field rises. One effect is distortion of the body. If the astronaut falls in feet-first, the feet will be closer than the head to the center of the hole, where gravity is stronger. As a result, the astronaut's feet will be pulled downward harder, stretching the body lengthwise. At the same time, the shoulders will be pulled toward the center of the hole on converging paths, so the astronaut will be squeezed sidewise. This stretching and squeezing process is sometimes referred to as spaghettification.

Theory suggests that at the center of the black hole, gravity rises without limit. Because the gravitational field manifests itself as a curvature, or warping, of spacetime, the escalating gravity is accompanied by a spacetime warp that also rises without known limit. Mathematicians refer to this feature as a spacetime singularity. It represents a boundary, or edge, of space and time through which the normal concept of spacetime cannot be continued. Many physicists believe that the singularity inside a black hole genuinely represents the end of space and time, and that any matter that encounters it will be completely obliterated. If this is the case, then even the atoms of the astronaut's body will vanish into the singularity, in a nanosecond of ultraspaghettification.

If the black hole has a mass of ten million suns—similar to the hole that may lie at the center of the Milky Way—and is nonrotating, then the duration experienced by the astronaut in falling from the event horizon to the annihilating singularity will be about three minutes. Those last three minutes will be very uncomfortable; in practice, spaghettification will kill the hapless individual long before the singularity is reached. During this final phase, the astronaut will in any case be unable to see the fatal singularity, because light cannot escape from it. If the black hole in question is of just one solar mass, its radius is about three kilometers, and the journey from event horizon to singularity will occupy just a few microseconds.

Although the elapsed time to destruction is very swift as experienced in the falling astronaut's frame of reference, the hole's time warp is such that, viewed from afar, the astronaut's last journey appears to be in slow motion.

As the astronaut approaches the event horizon, the pace of events in the vicinity seem to the distant observer to get slower and slower. In fact, it seems that it must take an infinite length of time for the astronaut to reach the horizon. So what amounts to eternity in the faraway regions of the universe is experienced all in a rush by the astronaut. In this respect, a black hole is a sort of gateway to the end of the universe, a cosmic blind alley representing an exit to nowhere. A black hole is a little region of space that contains the end of time. By jumping into one, those who may be curious about the end of the universe can experience it directly for themselves.

Although gravity is by far the weakest force of nature, its insidious and cumulative action serves to determine the ultimate fate not only of individual astronomical objects but of the entire cosmos. The same remorseless attraction that crushes a star operates on a much grander scale on the universe as a whole. The outcome of this universal attraction depends delicately on the total amount of matter there is to exert a gravitational pull. To find that out, we have to weigh the universe.

· ·

CHAPTER 6

WEIGHING THE UNIVERSE

It is often said that what goes up must come down. The tug of gravity on a body projected skyward acts to brake its flight and pull it back to Earth. But not always. If the body moves fast enough, it can escape the Earth's gravity altogether and fly off into space, never to return. Rockets that launch planetary spacecraft manage to achieve these high speeds.

The critical "escape velocity" is about 11 kilometers per second (25,000 miles per hour)—more than twenty times as fast as the Concorde. This critical number derives from both the mass of the Earth—that is, the amount of matter it contains—and its radius. The smaller a body of given mass is, the larger will be its surface gravity. To escape the solar system means overcoming the sun's gravity; the required escape velocity is 618 kilometers per second. Escaping from the Milky Way also requires a velocity of a few hundred kilometers per second. At the other extreme, the velocity required to escape from a compact object like a neutron star is tens of thousands of kilometers per second, while that for a black hole is the speed of light (300,000 kilometers per second).

What about escaping from the universe? As I pointed out in chapter 2, the universe does not appear to have an edge to escape from, but if we pretend that it does, and that the edge is situated at the limit of our observations (about fifteen billion light-years away), then the escape velocity will be about the velocity of light. This is a very significant result, because the most distant galaxies appear to be receding from us at close to the speed of light. Taken at face value, the galaxies seem to be flying apart so fast that they may indeed just "escape" from the universe, or at least from one another, and "never come down."

In fact, it turns out that the expanding universe behaves in a manner closely analogous to a body projected from Earth, even if there is no well-defined edge. If the rate of expansion is fast enough, the retreating galaxies will escape from the cumulative gravity of all the other material in the universe, and the expansion will continue forever. On the other hand, if the rate is too slow, the expansion will eventually be brought to a halt and the universe will start to contract. The galaxies will then "come down" again, and the ultimate cosmic catastrophe will ensue, as the universe collapses.

Which of these scenarios will come to pass? The answer depends on a comparison of two numbers. On the one hand, there is the rate of expansion; on the other, there is the total gravitational pull of the universe—in effect, the weight of the universe. The bigger the pull, the faster the universe must expand to overcome it. Astronomers can measure the rate of expansion directly by observing the redshift effect; however, there is still some controversy over the answer. The second quantity—the weight of the universe—is even more problematical.

How do you weigh the universe? It seems a daunting task; clearly we cannot do it directly. Nevertheless, we might be able to deduce its weight using the theory of gravitation. A lower limit is straightforward to attain. It is possible to weigh the sun by measuring its gravitational pull on the planets. We know that the Milky Way contains about a hundred billion stars of roughly one solar mass on average, so this provides a crude lower limit to the mass of the galaxy. We can now tot up how many galaxies there are in the universe. You can't add them individually—there are too many—but a good guesstimate is ten billion. This comes to 10^{21} solar masses, or about 10^{48} tons in all. Taking the radius of this assemblage of galaxies to be fifteen billion light-years, we can calculate a minimum value for the escape velocity from the universe: the answer turns out to be about 1 percent of the velocity of light. We can conclude that if the weight of the universe were due only to the stars the universe would escape its own gravitational pull and go on expanding indefinitely.

That is indeed what many scientists believe will happen. But not all astronomers and cosmologists are convinced that the sums have been done correctly. The matter we see is less than what is actually there, because not all objects in the universe shine. Dark bodies, such as dim stars, planets, and black holes, largely escape our attention. There is also a lot of dust and gas, much of it inconspicuous. In addition, the spaces between the galaxies are undoubtedly not entirely devoid of matter: there may be large quantities of tenuous gas there.

A more intriguing possibility, though, has excited astronomers for several years. The big bang, in which the universe originated, was the source of all the matter we see but also the source of much matter we don't see. If the universe began as an intensely hot soup of subatomic particles, then in addition to the familiar electrons, protons, and neutrons that make up ordinary matter, all sorts of other particles—recently identified in the laboratory by particle physicists—must also have been created in copious quantities. Most of these other types of particles are highly unstable, and would soon have decayed, but some may persist to the present epoch as relics of the cosmic origin.

Chief among the relics of interest are neutrinos, those ghostly particles whose activity is revealed in supernovas (see chapter 4). As far as we know, neutrinos cannot decay into anything else. (There are actually three different types of neutrinos, and they may be able to change into each other, but I shall ignore this complication here.) So we expect the universe to be bathed in a sea of cosmic neutrinos left over from the big bang. Assuming the energy of the primeval universe was shared democratically among all subatomic species, it is possible to calculate how many cosmic neutrinos there should be. The answer works out to be about a million neutrinos per cubic centimeter of space—or about a billion neutrinos for every particle of ordinary matter.

I have always been fascinated by this remarkable conclusion. At any given time, there are about one hundred billion neutrinos inside your body, almost all relics of the big bang, left more or less undisturbed since the first millisecond of existence. Because neutrinos move at or close to the speed of light, they zip through you so fast that every second you are penetrated by one hundred billion billion of them! This ceaseless violation goes entirely unnoticed by us, because neutrinos interact so weakly with ordinary matter that there is negligible probability that even one of them will be stopped inside your body in your lifetime. Nevertheless, the existence of so many neutrinos spread throughout the seemingly empty spaces of the universe could have profound consequences for its ultimate fate.

Although neutrinos are exceedingly weakly interacting, they do exert a gravitational force in common with all particles. They may not often significantly push and pull other matter around, but their indirect gravitational effects could prove crucial by adding to the total weight of the universe. To determine how much the neutrinos contribute it is necessary to know their mass.

Where gravity is concerned, it is the actual mass rather than the rest mass which counts. Because neutrinos move close to the speed of light, they may have a significant mass even though their rest mass is tiny (see page 39). Indeed, they may even have zero rest mass and move precisely at the speed of light. If this is so, then their actual mass can be determined by reference to their energy, which in the case of relic cosmic neutrinos can be deduced from their assumed energy acquired from the big bang. This original energy must be corrected by a factor that allows for the debilitating effect of the expansion of the universe. When all this is done, it turns out that neutrinos with zero rest mass would make no significant contribution to the total weight of the universe.

On the other hand, we can't be sure that the neutrino *does* have zero rest mass, nor that all three species of neutrino have the same rest mass. Our present theoretical understanding of neutrinos does not rule out a finite rest mass, and so it becomes a matter of experiment to determine what is the case. As mentioned in chapter 4, we know that if the neutrino does have a rest mass, it is certainly very small—much smaller than the rest mass of any other known particle. However, because there are so many neutrinos in the universe, even a tiny rest mass could make a big difference to the universe's total weight. It is a finely balanced affair. A mass as small as one ten-thousandth of the mass of the electron (otherwise the lightest-known particle) would be enough to make a dramatic impact: the neutrinos would then outweigh all the stars.

Detecting a rest mass as small as this is very difficult, and the results of experiments have been confusing and contradictory. Curiously, the detection of neutrinos from Supernova 1987A provided an important clue. As already remarked, if neutrinos have zero rest mass they must all travel at exactly the same speed—the speed of light. On the other hand, if the neutrino has a small but nonzero rest mass, then a range of speeds is possible. Neutrinos from a supernova are likely to be very energetic and therefore to move at very close to the speed of light even if they do have a nonzero rest mass. However, because they will have traveled through space for a long time, tiny variations in speed could translate into measurable variations in arrival time at Earth. By studying the extent to which the neutrinos from Supernova 1987A were spread over time, an upper limit can be placed on their rest mass of about one thirty-thousandth of the mass of the electron.

Unfortunately, the situation is further complicated, because there is known to be more than one type of neutrino. Most of the determinations of rest mass refer to the neutrino originally postulated by Pauli, but since its discovery a second type of neutrino has been found and the existence of a third type inferred. All three species would have been created in abundance in the big bang. It is very difficult to place limits directly on the mass of the other two neutrino types. Experimentally, the range of possible values remains very wide, but current thinking among theorists is that neutrinos probably do not dominate the mass of the universe. This sentiment could easily be reversed, in the light of new experimental determinations of neutrino masses.

Nor are neutrinos the only possible cosmic relics to consider when it comes to estimating the weight of the universe. Other stable, weakly interacting particles could have been created by the big bang, perhaps with rather larger masses. (If the rest mass becomes too large, their production is suppressed relative to other, lower-mass particles, because more energy is required to produce them.) Collectively these are known as WIMPs, for Weakly Interacting Massive Particles. Theorists have quite a shopping list of hypothetical WIMPs, bearing outlandish names like gravitinos, Higgsinos, and photinos. Nobody knows if they really exist, but if they do they will have to be taken into account in determining the weight of the universe.

Remarkably, it may be possible to test for the existence of WIMPs directly, from the way they are assumed to interact with ordinary matter. Although this interaction is predicted to be very weak, the large mass of the WIMPs enables them to pack a lot of punch. Experiments in a salt mine in the northeast of England and beneath a dam near San Francisco have been planned to spot passing WIMPs. Assuming the universe is replete with them, there would be an enormous number of WIMPs going through us (and the Earth) all the time. The principle of the experiment is mind-boggling: to detect the *sound* a WIMP makes when it bangs into an atomic nucleus!

The apparatus consists of a crystal of germanium or silicon surrounded by a cooling system. If a WIMP strikes a nucleus in the crystal, its momentum will cause the nucleus to recoil. This sudden shock creates a tiny sound wave, or vibration, in the crystal lattice. As the wave spreads, it will be damped and turned into heat energy. The experiment is designed to detect the minute pulse of heat associated with the decaying sound wave. Because the crystal is cooled to near absolute zero, the detector is extremely sensitive to the injection of any heat energy.

Theorists conjecture that galaxies are immersed in blob-shaped swarms of rather slow-moving WIMPs, with masses that could lie anywhere between one and a thousand proton masses and typical speeds of a few thousand kilometers per second. As our solar system orbits the galaxy, it sweeps through this invisible sea, and every kilogram of matter on Earth could scatter as many as a thousand WIMPs per day. Given this event rate, direct detection of WIMPs ought to be feasible.

While the hunt for WIMPs continues, the problem of weighing the universe is also being tackled by astronomers. Even if a body cannot be seen (or heard), the effects of its gravitational pull can still be apparent. For example, the planet Neptune was discovered because astronomers noticed that Uranus's orbit was being perturbed by the gravitational force of an unknown body. The dim white dwarf star Sirius B, which circles the bright star Sirius, was also discovered this way. Thus, by monitoring the motion of visible objects, astronomers can build up a picture of unseen matter too. (I have already explained how this technique has led to the suspicion that there may be a black hole in Cygnus X-1.)

Over the last decade or two, careful studies have been made of the way in which the stars in our galaxy move. Stars orbit the center of the Milky Way on a time scale typically in excess of two hundred million years. The galaxy is shaped rather like a disk, with a large blob of stars near the center. There is thus a crude resemblance to the solar system, in which planets orbit the sun; but there the inner planets, such as Mercury and Venus, move faster than the outer planets, such as Uranus and Neptune, because the inner planets feel a stronger gravitational pull from the sun. You might expect this rule to apply to the galaxy too: the stars near the periphery of the disk should move much more slowly than those near the center.

The observations, however, contradict this. Stars move at about the same speed throughout the disk. The explanation must be that the mass of the galaxy is not concentrated near the middle but is spread out more or less evenly. The fact that the galaxy *looks* as if it is concentrated near the middle suggests that the luminous material is only part of the story. Evidently there is a lot of dark or invisible material present, much of it in the outer reaches of the disk, accelerating the stars in that region. There could even be substantial amounts of dark matter beyond the visible edge and out of the plane of the luminous disk altogether, enveloping the Milky Way in an invisible massive halo that extends far into intergalactic space. A similar pattern of motion is observed in other galaxies. Measurements indi-' cate that the visible regions of galaxies are, on average, more than ten times as massive as their brightness (by comparison with the sun) might suggest, this ratio rising as high as five thousand times in the outermost regions.

The same sort of conclusion follows from the study of the motions of galaxies within galactic clusters. Clearly, if a galaxy moves fast enough it will escape the gravitational pull of the cluster. If all the galaxies in the cluster move this fast, the cluster will soon break up. A typical cluster of several hundred galaxies is situated in the constellation of Coma, and has been studied intensively. The average speed of the Coma galaxies is far too high for the cluster to hold together, unless there is at least three hundred times more mass present than can be accounted for by the luminous matter. Because it takes only a billion years or so for a typical galaxy to cross the Coma cluster, there has been plenty of time for the cluster to disperse by now. Yet that has not happened, and the structure of the cluster gives every impression of being gravitationally bound. Some form of dark matter seems to be present in substantial quantities, influencing the motion of the galaxies.

A further suggestion of unseen stuff comes from examining the very-large-scale structure of the universe—the way in which clusters and superclusters of galaxies clump together. As explained in chapter 3, galaxies are distributed in a manner reminiscent of froth, strung out in filaments or spread in huge sheets surrounding immense voids. Such a clumpy, frothy structure could not have arisen in the time available since the big bang without the added gravitational pull of nonluminous material. Computer simulations up to the time of writing cannot, however, reproduce the observed frothy structure with any simple form of dark matter, and it may be that a complicated cocktail is needed.

The latest scientific attention has focused on exotic subatomic particles as candidates for dark matter, but it could exist in more conventional forms, such as planetary-size masses or dim stars. Swarms of these dark objects might be wandering all around us in space and we would be blissfully unaware of the fact. Astronomers have recently discovered a technique that could reveal the existence of dark bodies that are not gravitationally bound to visible objects. The technique makes use of a result of Einstein's general theory of relativity known as gravitational lensing.

The idea is based on the fact that gravity can bend light rays. Einstein predicted that a starbeam passing close to the sun will be slightly curved, thereby displacing the apparent position of the star in the sky. By comparing the star's position with and without the sun in its vicinity, the prediction can be tested. This was first done by the British astronomer Sir Arthur Eddington in 1919, and brilliantly confirmed Einstein's theory.

Lenses bend light rays too, and as a result they can focus the light to form an image. If a massive body is symmetric enough, it can mimic a lens and focus light from a distant source. Figure 6.1 shows how. Light from a source S falls on a spherical body, and the gravity of the body curves the light around it, directing it to a focal point on the far side. The bending effect is tiny for most objects, but over astronomical distances even a slight curvature in the path of the light will eventually produce a focus. If the body interposes itself between Earth and the distant source S, the effect will appear as a greatly brightened image of S or, in exceptional cases where the line of sight is exact, as a bright circle of light known as an Einstein ring. For bodies with more complicated forms, the lensing will most likely produce multiple images rather than a single focused image. Astronomers have discovered a

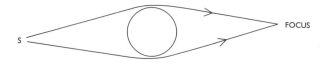

FIGURE 6.1

Gravitational lens. The gravity of the massive body (blob) bends the light rays from the distant source S. In a favorable case, this would produce a focusing effect. An observer at the focus would see a ring of light around the body.

• • • • • • • • •

number of gravitational lenses on a cosmological scale: galaxies in near-perfect alignment between Earth and distant quasars produce multiple images of these quasars, and in some cases arcs and entire rings of quasar light.

In their search for dark planets and dim dwarf stars, astronomers look for the tell-tale signs of lensing that would occur if such a body were interposed directly between Earth and a star. The image of the star would rise and fall in brightness in a distinctive manner as the dark body moved across the line of sight. Although the body itself would remain invisible, its presence would be inferred from the lensing effect. Some astronomers are using this technique to search for dark objects in the halo of the Milky Way. Although the probability of an exact alignment with a distant star is incredibly small, gravitational lensing ought to be observed if there are enough dark objects out there. In late 1993, a joint Australian-American team observing stars in the Large Magellanic Cloud from the Mt. Stromlo observatory, in New South Wales, reported what appears to be the first definite example of gravitational lensing by a dwarf star in the halo of our galaxy.

Black holes will also act as gravitational lenses, and extensive searches have been made for them using extragalactic radio sources (radio waves are lensed in the same way as light waves). Very few likely candidates have been found, leading to the impression that stellar or galactic-mass black holes are unlikely to account for much of the dark matter.

Not all black holes would show up in a lensing survey, however. It is possible that the extreme conditions prevailing shortly after the big bang encouraged the formation of microscopic black holes, perhaps no larger than an atomic nucleus. Such objects would have a mass equivalent to that of an asteroid. A lot of mass could be very effectively hidden in this form, and would be spread throughout the universe. Surprisingly, it is possible to place observational limits even on these bizarre entities. The reason concerns something called the Hawking effect, which I shall explain properly in chapter 7. Briefly, microscopic black holes are likely to explode amid a shower of electrically charged particles. The explosion occurs after a definite time, which depends on the size of the hole: smaller holes explode sooner. A hole with the mass of an asteroid will explode after ten billion years, which means about now. One effect of such an explosion would be to create a sudden pulse of radio waves, so radio astronomers have checked. No likely pulses have been detected, and it has been calculated that therefore no more than one explosion can occur every three million years per cubic light-year of space. This means that no more than a tiny fraction of the mass of the universe is in the form of microscopic black holes.

Overall, estimates of the amount of dark matter in the universe vary from one astronomer to another. It is likely that dark matter outweighs luminous matter by at least ten to one, and figures of a hundred to one are sometimes quoted. It is an astonishing thought that astronomers don't know what most of the universe consists of. The stars that they had long supposed accounted for most of the universe turn out to make up a rather small portion of the total.

For cosmology, the crucial issue is whether there is enough dark matter to arrest the expansion of the universe. The minimum density of matter which just fails to halt the expansion is called the "critical density." Its value can be calculated to be about one hundred times the density of visible matter. Such an amount remains possible, though perhaps only just. It is to be hoped that the search for dark matter will soon provide a definite yes or no, for nothing less than the ultimate fate of the universe hangs on it.

Given our present state of knowledge, we cannot say whether the universe will expand forever or not. If it will eventually start to contract, the question arises of when this will happen. The answer depends on precisely by how much the weight of the universe exceeds the critical weight. If it is 1 percent more than the critical weight, the universe will start contracting in about a trillion years; if it is 10 percent more, contraction is hastened to one hundred billion years from now.

Meanwhile, some theorists believe that it may be possible to establish the weight of the universe by calculation alone, without the need for difficult direct observations. The belief that human beings might divine profound cosmological knowledge merely from the power of reasoning follows a tradition stretching back to the ancient Greek philosophers. In the scientific era, a number of cosmologists have tried to formulate mathematical schemes that would yield the mass of the universe as a quantity fixed in value by some deep set of principles. Especially beguiling are those systems in which the exact number of particles in the universe is determined in terms of some numerological formula. Such largely armchair musings have not commended themselves to the majority of scientists, fascinating though they may be. In recent years, however, a more convincing theory that makes a definite prediction about the mass of the universe has become popular. This is the inflationary scenario discussed in chapter 3.

One of the predictions of the inflationary theory concerns the amount of matter in the universe. Suppose the universe starts out with a mass density much greater or less than the critical value at which collapse just fails to occur. When the universe embarks on the inflationary phase, the density changes dramatically, and in fact the theory predicts that it rapidly approaches the critical density. The longer the universe inflates, the closer the density gets to criticality. In the standard version of the theory, inflation lasts for only a very brief duration, so unless by a miracle the universe began with exactly the critical density, it will emerge from the inflationary phase with a density slightly greater or less than criticality.

However, the approach to the critical density during inflation occurs *exponentially* fast, so that the final value of the density is likely to be exceedingly close to the critical value, even for inflationary periods lasting just a tiny fraction of a second. The meaning of "exponential" here is that for roughly each extra tick that inflation persists the time that will elapse between the big bang and the onset of contraction *doubles*. So if, say, one hundred ticks' worth of inflation leads to contraction one hundred billion years later, then one hundred and one ticks implies contraction *two* hundred billion years later, while one hundred and ten ticks corresponds to contraction one hundred trillion years later. And so on.

How long did inflation last? Nobody knows, but for the theory to successfully explain the numerous cosmological puzzles I have described, it must endure for a certain minimum number of ticks (roughly one hundred; the figure is rather elastic). However, there is no *upper* limit. If by some extraordinary coincidence the universe inflated by only the minumum needed to explain our current observations, then the density after inflation could still be significantly above (or below) the critical value—in which case forthcoming observations should be able to determine the epoch of contraction, or that there will be no contraction. Much more likely is that inflation continued for many more ticks than the minimum, resulting in a density very close indeed to the critical value. This means that if the universe *is* going to contract it won't do so for an enormous length of time yet—very many times the present age of the universe. If that is the case, human beings will *never know* the fate of the universe they inhabit.

....

CHAPTER 7

FOREVER IS A LONG TIME

The important thing about infinity is that it is not just a very big number. Infinity is qualitatively different from something that is merely stupendously, unimaginably huge. Suppose the universe were to continue expanding forever so that is has no end. For it to endure for all eternity means that it would have an infinite lifetime. If this were the case, any physical process, however slow or improbable, would *have* to happen sometime, just as a monkey forever tinkering on a typewriter would eventually type the works of William Shakespeare.

A good example is provided by the phenomenon of gravitational-wave emission, which I discussed in chapter 5. Only in the case of the most violent astronomical processes will energy loss in the form of gravitational radiation produce conspicuous changes. The emission of about a milliwatt caused by the Earth orbiting the sun has an infinitesimal effect on the Earth's motion. Yet even a milliwatt power drain, extended over trillions upon trillions of years, would eventually cause the Earth to spiral into the sun. Of course, it is likely to be engulfed by the sun long before this, but the point is that processes that are negligible on a human time scale, but are nevertheless persistent, may eventually come to predominate and thus serve to determine the ultimate fate of physical systems.

Let us imagine the state of the universe a very, very long time in the future—say, in a trillion trillion years. The stars have long since burned out; the universe is dark. But it is not empty. Amid the black vastness of space lurk spinning black holes, stray neutron stars, and black dwarfs—even a few planetary bodies. At this epoch, the density of such objects is exceedingly low: the universe has expanded to ten thousand trillion times its present size.

Gravity would play out a strange battle. The expanding universe attempts to pull every object farther apart from its neighbors, but the mutual gravitational attractions oppose this and try to bring bodies together. As a result, certain collections of bodies—for example, clusters of galaxies, or what pass for galaxies after eons of structural degeneration—remain gravitationally bound, but these collections drift ever farther from neighboring collections. The ultimate outcome of this tug-of-war depends on how fast the rate of expansion decelerates. The lower the density of matter in the universe, the more "encouragement" these collections of bodies get to disengage from their neighbors and move apart freely and independently.

Within a gravitationally bound system, the slow but inexorable processes of gravity exert their dominance. Gravitational-wave emission, feeble though it is, insidiously saps the system's energy, causing a slow spiral of death. Ever so gradually, dead stars creep closer to other dead stars or black holes, and coalesce in an extended orgy of cannibalism. It takes a trillion trillion years for gravitational waves to completely degrade the orbit of the sun, a black dwarf cinder silently gliding toward the galactic center, where a gigantic black hole waits to swallow it.

It is by no means certain, however, that the dead sun will meet its final demise in this way, for as it drifts slowly inward it will occasionally encounter other stars. Sometimes it will pass close to a binary system—a pair of stars locked in close gravitational embrace. The stage is then set for a curious phenomenon known as the gravitational slingshot. The motion of two bodies in orbit about each other displays a classic simplicity. It was this problem—in the guise of a planet orbiting the sun—that preoccupied Kepler and Newton and led to the birth of modern science. In an idealized situation, and ignoring gravitational radiation, the motion of the planet is regular and periodic. No matter how long you wait, the planet goes on orbiting just the same. The situation is dramatically different, however, if a third body is present—say, a star and two planets, or three stars. No longer is the motion simple and periodic. The pattern of mutual forces between the three bodies is always changing in a complicated way. As a result, the energy of the system is not shared equally among the participants, even if they are identical bodies. Instead, there is a complicated dance in which one body and then another has the lion's share of energy. Over long periods of time, the behavior of the system can be essentially random: in fact, the three-body problem of gravitational dynamics is a good example of a so-called chaotic system. It can happen that two of the bodies "gang up," conveying so much of the available energy to the third that it is kicked out of the system altogether, like a projectile from a sling. Hence the term "gravitational slingshot."

The slingshot mechanism can eject stars from star clusters, or even from the galaxy itself. In the far future, the great majority of dead stars, planets, and black holes will be flung off into intergalactic space this way—perhaps to encounter another disintegrating galaxy, or else to roam forever in the vast expanding emptiness. However, the process is slow: it will take a billion times the present age of the universe for this dissolution to be completed. The remaining few percent of objects will, by contrast, migrate to the centers of the galaxies and merge to form gigantic black holes.

As explained in chapter 5, astronomers have good evidence that there already exist monster black holes at the centers of some galaxies, greedily gobbling up swirling gases and releasing huge amounts of energy as a result. Such a feeding frenzy will await most galaxies in time, and will continue until the material surrounding the black hole has been either sucked up or ejected, perhaps to fall back again eventually or to join the dwindling intergalactic gases. The bloated black hole will then remain quiescent, with only the occasional rogue neutron star or small black hole plunging in. This will not, however, be the end of the black-hole story. In 1974, Stephen Hawking discovered that black holes are not completely black after all. Instead, they emit a feeble glow of heat radiation.

The Hawking effect can be properly understood only with the help of the quantum theory of fields, a difficult branch of physics that I have already alluded to in connection with the inflationary-universe theory. Recall that a central tenet of the quantum theory is Heisenberg's uncertainty principle, according to which quantum particles do not possess sharply defined values for all their attributes. For example, a photon or an electron cannot have a definite value for its energy at a specific moment of time. In effect, a subatomic particle can "borrow" energy, as long as it is paid back promptly.

As I noted in chapter 3, energy uncertainty leads to some curious effects, such as the fleeting presence in apparently empty space of short-lived, or virtual, particles. This leads to the strange concept of the "quantum vacuum"— a vacuum that, far from being vacuous and inert, seethes with restless virtual-particle activity. Although this activity usually goes unnoticed, it can produce physical effects. One such effect occurs when the vacuum activity is disturbed by the presence of a gravitational field.

An extreme case concerns the virtual particles that appear near the event horizon of a black hole. Recall that virtual particles live on borrowed energy for a very short time, after which the energy must be "paid back" and the particles obliged to disappear. If for any reason the virtual particles receive a big enough energy boost from some *external* source during their brief allotted time, the loan can be cleared on their behalf. There is then no longer any obligation for the particles to disappear to pay it off. The effect of this benefaction is therefore to promote the virtual particles to real particles, which are able to enjoy a more or less permanent existence.

According to Hawking, such a debt-clearing benefaction is what happens near a black hole. In this case, the "benefactor" that supplies the required energy is the gravitational field of the black hole. This is how the deal is struck: Virtual particles are usually created in pairs moving in opposite directions. Imagine such a pair of newly appeared particles just outside the event horizon. Suppose the motion of the particles is such that one of them falls across the horizon into the hole. It will pick up a great deal of energy from the hole's intense gravity as it goes. This energy boost, Hawking discovered, is enough to "clear the loan" completely and promote both the infalling particle and its partner—still residing outside the event horizon— to real particles. The fate of the deserted particle outside the horizon is a touch-and-go affair. It, too, may end up being sucked down the hole eventually, or it may fly off at high speed and escape from the black hole completely. Hawking thus predicts that there should be a steady flux of these escapees streaming away into space from the vicinity of the hole, constituting what is known as Hawking radiation.

The Hawking effect would be strongest for microscopic black holes. Because a virtual electron, for example, can travel at most about 10^{-11} centimeters under normal conditions before the loan is recalled, only black holes smaller than this (roughly, of nuclear dimensions) will effectively be able to create a stream of electrons. If the hole is any larger than this, most of the virtual electrons will not have time enough to get across the horizon before the loan has to be repaid.

The distance a virtual particle may traverse depends on how long it lives, which in turn is dictated—via the Heisenberg uncertainty principle—by the size of the energy loan. The bigger the loan, the shorter the life of the particle. A major component of the energy loan is the particle's rest-mass energy. In the case of an electron, the loan has to be at least equal to the electron's rest-mass energy. For a particle with a larger rest mass—for example, a proton—the loan would be bigger and hence briefer, so the distance traveled would be less. Therefore the production of protons by the Hawking effect requires a black hole even smaller than one of nuclear dimensions. Conversely, particles with a lower rest mass than electrons—for example, neutrinos—would be created by a black hole of greater than nuclear dimensions. Photons, which have *zero* rest mass, will be created by a black hole of any size. Even a black hole of one solar mass will have a Hawking flux of photons, and possibly neutrinos too; however, in such cases the intensity of the flux is very feeble.

Use of the word "feeble" here is no exaggeration. Hawking found that the spectrum of energy produced by a black hole is the same as that radiated from a hot body, so one way to express the strength of the Hawking effect is in terms of temperature. For a hole of nuclear size (10^{-13} centimeters in diameter), the temperature is very high—about ten billion degrees. By contrast, a black hole weighing one solar mass, which is over a kilometer in diameter, has a temperature of less than a ten-millionth of a degree above absolute zero. The entire object would emit no more than a billion-billion-billionth of a watt in Hawking radiation.

One of the oddities of the Hawking effect is that the temperature of the radiation goes up as the mass of the black hole goes down. This means that small holes are hotter than big ones. As a black hole radiates, it loses energy and hence mass, so it shrinks. Consequently, it gets hotter and radiates more vigorously, and therefore shrinks still faster. The process is inherently unstable and eventually runs away, with the black hole emitting energy and shrinking in size at an ever faster rate.

The Hawking effect predicts that eventually all black holes will simply disappear in a puff of radiation. The final moments would be spectacular, appearing like the detonation of a large nuclear bomb, a brief flash of intense heat energy followed by—nothing. At least that's what the theory suggests. But some physicists are unhappy that a material object can collapse to form a black hole, which in turn vanishes leaving only heat radiation. They worry that two very different objects could end up producing identical heat radiation, with no information about the original body surviving. Such a vanishing act violates all sorts of cherished conservation laws. An alternative proposal is that the disappearing hole leaves behind a minute remnant that somehow contains vast quantities of information. Either way, the overwhelming fraction of the hole's mass is radiated away in the form of heat and light.

The Hawking process is almost inconceivably slow. A black hole of one solar mass would take 10^{66} years to disappear, while a supermassive hole would take more like 10^{93} years. And the process wouldn't even get under way until the background temperature of the universe had dropped below that of the black hole, otherwise the heat flowing into the hole from the surrounding universe wins out over the heat flowing away from the hole via the Hawking effect. The cosmic background heat radiation left over from the big bang is currently at a temperature of about three degrees above absolute zero, and it would take

10^{22} years before this cooled to the level at which there would be a net heat loss from solar-mass black holes. The Hawking process isn't something you sit around and watch.

But forever is a long time, and given forever, eventually all black holes—even the supermassive ones—will probably disappear, their death pangs momentary flashes of light in the inky blackness of eternal cosmic night, a fleeting epitaph to the erstwhile existence of a billion blazing suns.

What's left?

Not all matter falls into black holes. We need to think about the neutron stars and black dwarfs and rogue planets that wander off alone into the vast intergalactic spaces, not to mention the tenuous gas and dust that never got itself together into stars, and the asteroids, comets, meteorites, and odd chunks of rock that clutter star systems. Do these things survive forever?

Here we run into theoretical difficulties. We need to know whether ordinary matter—the stuff of you and me and planet Earth—is absolutely stable. The ultimate key to the future lies with quantum mechanics. Although quantum processes are normally associated with atomic and subatomic systems, the laws of quantum physics should apply to everything, including macroscopic bodies. Quantum effects in large objects are exceedingly tiny, but over vast periods of time they would be able to bring about major changes.

The hallmarks of quantum physics are uncertainty and probability. In the quantum realm nothing is certain, except the betting odds. This means that if a process is at all possible, given enough time it will occur eventually, however improbable it may be. We can observe this rule at work in the case of radioactivity. A nucleus of uranium-238 is almost completely stable. There is, however, a minute chance that it will eject an alpha particle and transmute into thorium. To be precise, there is a certain very small probability per unit time that a given uranium nucleus will decay. On average, it takes about four and a half billion years to happen, but because the laws of physics demand a fixed probability per unit time, any *given* uranium nucleus is certain to decay eventually.

Radioactive alpha decay takes place because there is a small uncertainty in the location of the protons and neutrons that make up the nucleus of a uranium atom, so there is always a tiny probability that a cluster of these particles will be momentarily located *outside* the nucleus, whence they are rapidly propelled away. Likewise, there is an even smaller, but still nonzero, uncertainty in the precise position of an atom in a solid. For example, an atom of carbon in a diamond will reside in a very well-defined location in the crystal lattice, and at the near-zero temperatures expected in the far future of the universe this residence would be exceedingly stable. But not completely. There is always a tiny uncertainty in the position of the atom, which implies a tiny probability that the atom may spontaneously jump out of its place in the lattice and appear somewhere else. Because of such migratory processes, nothing—not even a substance as hard as diamond—is truly solid. Instead, apparently solid matter is like an exceedingly viscous liquid, and over immense durations it can flow, due to quantum-mechanical effects. The theoretical physicist Freeman Dyson has estimated that after about 10^{65} years not only would every carefully cut diamond be reduced to a spherical bead but every chunk of rock would likewise deform into a smooth ball.

Position uncertainty could even lead to nuclear transmutations. For example, consider two neighboring atoms of carbon in the diamond crystal. Very rarely, the spontaneous relocation of one such atom will cause its nucleus to appear, momentarily, right next to the nucleus of its neighbor. Nuclear attractive forces may then cause the two nuclei to fuse to form a nucleus of magnesium. So nuclear fusion doesn't require very high temperatures: cold fusion *is* possible, but it takes a stupendous length of time. Dyson has estimated that after 10^{1500} years (that is, 1 followed by fifteen hundred zeros!), all matter will transmute in this manner into the most stable nuclear form, which is the element iron.

However, it may be that nuclear matter will not survive this long anyway, due to more rapid, but still incredibly slow, transmutation processes. Dyson's estimate assumes that protons (and neutrons bound in nuclei) are *absolutely* stable. In other words, if a proton doesn't fall into a black hole and is left otherwise undisturbed, it will last forever. But can we be sure this is so? When I was a student, nobody doubted it. Protons were forever. They were supposed to be completely stable particles. But there was always a nagging doubt about this. The problem concerns the existence of a particle called the positron, which is identical to the electron except that, like the proton, it has a positive charge. Positrons are much lighter than protons, so, all else being equal, protons would prefer to transmute into positrons: it is a deep principle of physics that physical systems seek out their lowest energy state, and low mass means low energy. Now, nobody could say why protons don't simply go ahead and do this, so physicists simply assumed that there was a law of nature which forbade it. Until recently, this topic was not at all well understood, but in the late 1970s a clearer picture emerged concerning the way in which nuclear forces prompt particles to transmute into one another quantum-mechanically. The latest theories have a natural place for the law that forbids proton decay, but most of these theories also predict that the law is not 100 percent effective. There could be a *very* tiny probability that a given proton will indeed transmute into a positron. The leftover mass is predicted to appear partly in the form of an electrically neutral particle, such as a so-called pion, and partly in the form of energy of motion (the decay products would be created moving at high speed).

In one of the simplest theoretical models, the average time required for a proton to decay is 10^{28} years, which is a billion billion times longer than the present age of the universe. You might therefore imagine that the subject of proton decay would remain a purely academic curiosity. However, it must be remembered that the process is quantum-mechanical, and hence inherently probabilistic in nature: 10^{28} years is the predicted *average* lifetime, not the *actual* lifetime for every proton. Given enough protons, there is a good chance that one will decay before your very eyes. In fact, given 10^{28} protons, you might expect roughly one decay per year, and 10^{28} protons are contained in a mere 10 kilograms of matter.

As it happens, a proton lifetime of this duration had already been ruled out by experiment before the theory was popular. However, different versions of the theory gave longer lifetimes—10^{30} or 10^{32} years, or even longer (some theories predict as long as 10^{80} years). The lower values lie on the edge of experimental testability. A decay time of 10^{32} years, for example, would mean that you might lose one or two protons from your body this way during your lifetime. But how to detect such rare events?

The technique adopted was to assemble thousands of tons of matter and monitor it for many months with sensitive detectors tuned to be triggered by the products of a proton-decay event. Unfortunately, the search for proton decay is of a needle-in-a-haystack nature, because such decays are masked by a much greater number of similar events caused by the products of cosmic radiation. The Earth is continually bombarded by high-energy particles from space, which produce an ever-present background of subatomic debris. To reduce this interference, the experiments need to be conducted deep underground.

One such experiment was set up more than half a mile below ground, in a salt mine near Cleveland, Ohio. The apparatus consisted of 10,000 tons of ultrapure water in a cubical tank surrounded by detectors. Water was chosen on account of its transparency, enabling the detectors to "see" as many protons at once as possible. The idea was this: If a proton decays in the manner expected from the fashionable theories, then it produces, as explained, an electrically neutral pion in addition to a positron. The pion in turn rapidly decays, usually into two very energetic photons, or gamma rays. Finally, these gamma rays encounter nuclei in the water and each creates an electron-positron pair, also very energetic. In fact, these secondary electrons and positrons would be so energetic that they would travel at close to the speed of light, even in the water.

Light travels at 300,000 kilometers per second in a vacuum, and this is the limiting speed at which *any* particle can travel. Now water has the effect of slowing light down somewhat, to roughly 230,000 kilometers per second. Therefore, a high-speed subatomic particle moving at nearly 300,000 kilometers per second through water actually travels faster than light travels *in water*. When an aircraft travels faster than sound, it creates a sonic boom. Similarly, a charged particle that travels through a medium faster than light travels in that medium creates a distinctive electromagnetic shock wave—known as Cerenkov radiation, after its Russian discoverer. So the Ohio experimenters set up a collection of light-sensitive detectors to search for Cerenkov flashes. In order to distinguish proton-decay events from cosmic neutrinos and other spurious subatomic debris, the experimenters looked for a distinctive signature—back-to-back simultaneous pairs of Cerenkov light pulses, which would be emitted by the oppositely moving electron-positron pairs.

Unfortunately, after several years of operation the Ohio

experiment failed to find convincing evidence for proton decay—although, as noted in chapter 4, it did pick up the neutrinos from Supernova 1987A. (As so often in science, looking for one thing leads to the unexpected discovery of another.) Other experiments, using different designs, have also led to null results at the time of writing. This may mean that protons do not decay. On the other hand, it may mean that they do decay but that their lifetime is in excess of 10^{32} years. To measure a decay rate slower than this is beyond current experimental possibility, so probably the jury will remain out on proton decay for the foreseeable future.

The search for proton decay was stimulated by theoretical work on the various grand unified theories, which set as their goal the unification of the strong nuclear force (the force that binds protons and neutrons together in nuclei) with the weak nuclear force (responsible for beta radioactivity) and the electromagnetic force. Proton decay would occur as a result of a minute intermingling of these forces. But even if this grand-unification idea turns out to be wrong, there remains the possibility that protons will decay via another route—one that involves the fourth fundamental force of nature, gravity.

To see how gravity can cause proton decay, it is necessary to take into account the fact that the proton is not a truly elementary particle with a pointlike form. It is actually a composite body made up of three smaller particles called quarks. Most of the time, the proton has a diameter of about a ten-trillionth of a centimeter, this being the average distance between the quarks. However, the quarks do not remain at rest but are continually changing their positions inside the proton, because of quantum-mechanical uncertainty. From time to time, two quarks will approach each other very closely. Still more rarely, all three quarks will find themselves in extremely close proximity. It is possible that the quarks will get *so* close that the gravitational force between them, normally utterly negligible, will overwhelm all else. If this happens, the quarks will fall together to make a minuscule black hole. In effect, the proton collapses under its own gravity by quantum-mechanical tunneling. The resulting minihole is highly unstable—recall the Hawking process—and more or less instantly vanishes, creating a positron. Estimates of the lifetime for proton decay via this route are very uncertain, and vary from 10^{45} years to a stupendous 10^{220} years.

If protons do decay after an immense duration, the consequences for the far future of the universe are profound. All matter would be unstable, and would eventually disappear. Solid objects, like planets, that had avoided falling into a black hole would not last forever. Instead, they would very gradually evaporate. A proton lifetime of, say, 10^{32} years would imply that the Earth is losing a trillion protons every second. At this rate, after 10^{33} years or so our planet would effectively have vanished, assuming that something else hadn't destroyed it already.

Neutron stars are not immune from this process. Neutrons are also made up of three quarks, and can transmute into lighter particles by mechanisms similar to those that spell the demise of protons. (Isolated neutrons are in any case unstable, and decay within about fifteen minutes.) White dwarf stars, rocks, dust, comets, tenuous clouds of gas, and all the other astronomical paraphernalia would likewise succumb in the fullness of time. The 10^{48} tons of ordinary matter that we presently observe spread throughout the universe is all destined to disappear either into black holes or through slow nuclear decay.

Of course, when protons and neutrons decay, they create decay products, so the universe is not necessarily left entirely devoid of matter of any sort. For example, as already mentioned, one likely decay route for a proton is into a positron plus a neutral pion. The pion is very unstable and promptly decays into two photons, or perhaps into an electron-positron pair. Whichever is the case, the universe will gradually acquire more and more positrons as a result of proton decay. Physicists believe that the total number of positively charged particles (currently mainly protons) in the universe is the same as the number of negatively charged particles (mainly electrons). This implies that, once all the protons have decayed, there will be an equal mixture of electrons and positrons. Now, the positron is the so-called antiparticle of the electron, and if a positron meets an electron they annihilate each other—a process readily studied in the laboratory—releasing energy in the form of photons.

Calculations have been carried out to try to determine whether the positrons and electrons left in the far future of the universe will annihilate each other completely, or whether there will always remain a small residue. Annihilation does not take place abruptly. Instead, an electron and a positron first arrange themselves into a sort of mini-atom called positronium, with both particles orbiting their common center of mass in a dance of death, bound by their mutual electric attraction. The particles then spiral together and annihilate. The time taken to spiral in depends on the initial distance between the electron and the positron when the positronium "atom" forms. In the laboratory, positronium decay takes place in a tiny fraction of a second, but in outer space, with little to disturb them, electrons and positrons could become bound in a huge orbit. Estimates indicate that it would take 10^{71} years for most electrons and positrons to form positronium, but in most of these cases their orbits would be many trillions of light-years in diameter! The particles would move so slowly that they would take a million years to travel a centimeter. So sluggish would the electrons and positrons have become that the spiral time works out at a staggering 10^{116} years. Nevertheless, the final fate of these positronium atoms is sealed from the moment they form.

Curiously, not all electrons and positrons need annihilate. All the while that electrons and positrons seek out their opposite numbers, the density of these particles steadily declines, both as a result of annihilation and also because of the continuing expansion of the universe. As time goes on, it gets harder and harder for positronium to form. So although the tiny residue of remaining matter gets less and less, at no time does it ever disappear completely. There will always be the odd electron or positron to be found somewhere, even though each such particle resides in solitude within an ever-growing volume of empty space.

We can now paint a picture of what the universe would be like after all these incredibly slow processes have been completed. First, there will be the stuff left over from the big bang, the cosmic background that has been there all along. This consists of photons and neutrinos, and maybe some other completely stable particles we don't yet know about. The energy of these particles will go on declining as the universe expands, until they form a totally negligible background. The ordinary matter of the universe will have disappeared. All the black holes will have evaporated. Most of the mass of the black holes will have gone into photons, though some will also be in the form of neutrinos, and a very tiny fraction, emitted during the final explosive burst of the holes, will be in the form of electrons, protons, neutrons, and heavier particles. The heavier particles all rapidly decay, and the neutrons and protons decay more slowly, leaving a few electrons and positrons to join those others that are the last remaining residue of the ordinary matter we see today.

The universe of the very far future would thus be an inconceivably dilute soup of photons, neutrinos, and a dwindling number of electrons and positrons, all slowly moving farther and farther apart. As far as we know, no further basic physical processes would ever happen. No significant event would occur to interrupt the bleak sterility of a universe that has run its course yet still faces eternal life—perhaps eternal death would be a better description.

This dismal image of cold, dark, featureless near-nothingness is the closest that modern cosmology comes to the "heat death" of nineteenth-century physics. The time taken for the universe to degenerate to this state is so long that it defies human imagination. Yet it is but an infinitesimal portion of the infinite time available. As remarked, forever is a long time.

Although the decay of the universe occupies a duration so vastly in excess of human time scales that it is virtually meaningless to us, people are still eager to ask, "What will happen to our descendants? Are they inevitably doomed by a universe that will slowly but inexorably shut down around them?" Given the rather unpromising state that science predicts for the universe of the far future, it seems that any form of life must ultimately be doomed. But death is not that simple.

....

CHAPTER 8

LIFE IN THE SLOW LANE

In 1972, an organization called the Club of Rome published a gloomy prognosis about the future of humanity called *The Limits to Growth*. Among their many claims of imminent disaster was the prediction that the world's supply of fossil fuels would run out within a very few decades. People became alarmed, oil prices shot up, and alternative energy research became fashionable. Here we are in the 1990s, and there is no sign yet that fossil fuels are on the point of exhaustion. As a result, alarm has been replaced by complacency. Unfortunately, simple arithmetic dictates that a finite resource cannot go on being depleted at a nondiminishing finite rate forever. Sooner or later the energy crunch will be upon us. A similar conclusion can be drawn concerning the Earth's population: it cannot go on growing indefinitely.

Some Jeremiahs believe that the ensuing energy and overpopulation crises will see off humanity once and for all. There is no need, though, to draw a parallel between the disappearance of fossil fuels and the disappearance of *Homo sapiens*. There are vast sources of energy all about us, if only we have the will and ingenuity to harness them. Most notably, sunlight has more than enough energy for our purposes. A tougher problem is to curb population growth before massive starvation does it for us. This requires social, economic, and political skills rather than scientific ones. However, *if* we can overcome the energy bottleneck occasioned by the depletion of fossil fuels, and *if* we can stabilize the human population without disastrous conflict, and *if* the ecological and asteroid-impact damage to the planet can be limited, then I believe that humanity is set to flourish. There is no obvious law of nature that limits the longevity of our species.

In the previous chapters, I have described how, over mind-boggling durations of time, the structure of the universe will change—generally in the direction of degeneration—as a result of slow physical processes. Humans have been around for at most five million years (depending on the definition of human), and civilization (after a fashion) for merely a few thousand. The Earth could remain habitable for two or three *billion* years hence—with a limited population, of course. This is such an enormous time span that it beggars the imagination. It may seem so big as to be effectively infinite. Yet we have seen how even a billion years is the merest blip compared to the time scale for gross astronomical and cosmological change. Earthlike habitats may still exist elsewhere in our galaxy in a *billion* billion years.

We can certainly imagine our descendants, with such a vast amount of time at their disposal, developing space exploration and all manner of marvelous technologies. They will have plenty of time to leave Earth before the sun grills it to a crisp. They can seek out another suitable planet, and then another, and so on. By expanding into space, the population can expand too. Does this give comfort—to know that our struggle to survive in the twentieth century may not ultimately be in vain?

In chapter 2, I noted that Bertrand Russell, in a fit of depression over the consequences of the second law of thermodynamics, wrote in anguished terms about the futility of human existence given the fact that the solar system is doomed. Russell clearly felt that the apparently inevitable demise of our habitat somehow rendered human life pointless or even farcical. This belief certainly contributed to his atheism. Would Russell have felt better had he known that black-hole gravitational energy could outperform the sun many times and last for trillions of years after the solar system had disintegrated? Probably not. It is not the actual duration of time that counts but the idea that sooner or later the universe will become uninhabitable; this idea makes some people feel that our existence is pointless.

From the description given at the end of chapter 7 of the far future of the universe, it might be supposed that a less equable and more hostile environment can scarcely be imagined. However, we must not be chauvinistic, or pessimistic. Human beings would undoubtedly have a hard time making a living in a universe consisting of a dilute soup of electrons and positrons, but the important issue is surely not whether our species as such is immortal but whether our *descendants* can survive. And our descendants are unlikely to be human beings.

The species *Homo sapiens* emerged on Earth as a product of biological evolution. But the processes of evolution are rapidly being modified by our own activities. We have already interfered with the operation of natural selection. It is also becoming increasingly possible to engineer mutations. We may soon be able to design human beings with prescribed attributes and physical characteristics by direct genetic manipulation. These biotechnological opportunities have arisen in just a few decades of technological society. Imagine what could be achieved with thousands or even millions of years of science and technology.

In just a few decades, humankind has been able to leave the planet and venture into near space. Over the eons, our descendants could spread beyond Earth into the wider solar system, and then to other star systems within the galaxy. People often have the misconception that such an undertaking would take close to eternity. This is not so. Colonization would probably proceed by planet hopping: Colonists would leave Earth for a suitable planet a few light-years away, and if they could travel close to the speed of light the journey would take only those few years. Even if our descendants never achieve speeds more than 1 percent of the speed of light—a modest enough goal—then the travel time will be just a few centuries. The actual establishment of the new colony may take a few more centuries to complete, by which time the descendants of the original colonizers could think about dispatching their own colonizing expedition to yet another suitable planet farther on. After a another few hundred years, this next planet would be colonized, and so on. It was in this manner that the Polynesians colonized the islands of the central Pacific.

Light takes only about a hundred thousand years to cross the galaxy, so at 1 percent of this speed the total journey time is ten million years. If a hundred thousand planets are colonized along the way and each takes two centuries to become established, this does no more than treble the galactic-colonization time scale. But thirty million years is a very short time by astronomical or even geological standards. The Sun takes about two hundred million years to orbit the galaxy just once; life on Earth has existed for at least seventeen times as long as this. The Sun's aging will threaten Earth seriously only in two or three billion years, so in thirty million years very little change would have occurred. The conclusion is that our descendants could colonize the galaxy in a small fraction of the time that life on Earth took to evolve into a technological society.

What would these colonist descendants of ours be like? If we allow free rein to the imagination, we can conjecture that the colonists might be genetically engineered to adapt easily to the target planet. To take a simple example, if an Earthlike planet is discovered around the star Epsilon Eridani and found to have only 10 percent oxygen in its atmosphere, then colonists could be engineered to have more red blood corpuscles. If the surface gravity of the new planet is higher, they could be endowed with a more robust frame and stronger bones. And so on.

The journey need present no problem, either—even if it takes some centuries to complete. The spacecraft could be built as an ark—a completely self-contained ecosystem capable of supporting the travelers for many generations. Or the colonists could instead be deep-frozen for the journey. In fact, it would make more sense to send only a small craft and crew and include millions of frozen fertilized ova among the cargo. These could then be incubated on arrival, thus providing an instant population without the logistical and sociological problems of transporting large numbers of adult beings over a long duration.

Again, in the spirit of speculating what might be possible given enormous amounts of time, there is no reason why these colonists should be human in appearance or even in mentality. If beings can be engineered to meet various needs, then each expedition could involve purpose-designed entities with the necessary anatomy and psychology for the job.

The colonists need not even be living organisms, by the usual definition. Already it is possible to implant silicon-chip microprocessors into human beings. Further development of this technology could see a blend of organic and artificial electronic parts serving both physiological and brain functions. For example, it may be possible to design "bolt-on" memory for human brains, similar to those extra memories now available for computers. Conversely, it may soon prove more efficient to adapt organic material to perform computation than to produce solid-state devices for the job. In effect, it will be possible to "grow" computer components biologically. More likely, for many tasks digital computers will be replaced by neural nets; even now, neural nets are being used in place of digital computers to simulate human intelligence and predict economic behavior. And it may make better sense to grow organic neural nets from bits of brain tissue than to manufacture them *ab initio*. It may also be feasible to construct a symbiotic blend of organic and artificial networks. With the development of nanotechnology, the distinction between living and nonliving, natural and artificial, brain and computer, will become increasingly blurred.

At present, such speculations belong to the realm of science fiction. Can they become science fact? After all, just because we can imagine something does not mean that it will happen. However, we can apply the same principle to technological processes as we did to natural processes: given long enough, anything that *can* happen *will* happen. If humans or their descendants remain sufficiently motivated (that may be a big "if"), then technology will be bounded only by the laws of physics. A challenge like the human genome project, which may be a daunting task for a single generation of scientists, would be straightforward enough if a hundred, or a thousand, or a million generations arose to carry out the work.

Let us take the optimistic position that we will survive and continue to develop our technology toward its limits. What does that imply about the exploration of the universe? The construction of purpose-designed sentient beings would open up the possibility of sending agents into hitherto totally hostile habitats to perform currently unthinkable tasks. Although these beings may be the end products of human-initiated technology, they would not themselves be human.

Should we have a proprietary concern for the destiny of these weird entities? Many people may feel a sense of revulsion at the prospect of the replacement of humanity by such monsters. If survival requires human beings to give way to genetically engineered organic robots, perhaps we would opt for extinction. Yet if the likelihood of humanity's demise depresses us, we have to ask precisely what it is about human beings that we would wish to preserve. Surely not our physical form. Would it really disturb us to know that in, say, one million years from now, our descendants may have lost their toes? Or have shorter legs or bigger heads and brains? After all, our physical form has changed a lot anyway over the last few centuries, and there are currently wide variations among different ethnic groups.

When pressed, I suspect most of us would set more store by what might be called the human spirit—our culture, our set of values, our distinctive mental makeup, as exemplified in our artistic, scientific, and intellectual achievements. These things are certainly worth preserving and perpetuating. If we could pass on our essential *humanity* to our descendants, whatever their physical form, then survival of what matters most would be attained.

Whether it is possible to create humanlike beings who will go forth and spread across the cosmos is of course highly conjectural. Quite apart from anything else, it may be that humanity will lose the motivation for such a grand enterprise, or that economic, ecological, or other disasters will bring about our demise before we leave the planet in earnest. It may even be that extraterrestrial beings are one jump ahead of us and have already colonized most of the suitable planets (though evidently not Earth—yet). But whether the task falls to our descendants or to those of some alien species, the possibility of spreading across the universe and gaining control over it through technology is a fascinating one, and it is tempting to ask how such a superrace would contend with the slow degeneration of the universe.

The durations of time for physical decay discussed in chapter 7 are so enormous that any attempt to guess what technology may be like in the very far future based on extrapolating present trends on Earth are useless. Who can imagine a technological society one trillion years old? It might seem as if it could achieve anything. Nevertheless, any technology, however far advanced, would presumably still be subject to the basic laws of physics. If, for example, the theory of relativity is correct in its conclusion that no material body can exceed the speed of light, then even a trillion years of technological endeavor would fail to break the light barrier. More seriously, if all interesting activity involves using up at least some energy, then the continuing depletion of available free energy sources in the universe will eventually present a serious threat to a technological community, however advanced it may be.

By applying basic physical principles to the broadest definition of sentient beings, we can investigate whether the degeneration of the universe in the far future presents any truly fundamental obstacles to their survival. For a being to qualify for the description of "sentient" it must at least be able to process information. Thinking and experiencing are both examples of activities involving information processing. So what demands might this make on the physical state of the universe?

A characteristic feature of information processing is that it dissipates energy. This is the reason the word processor on which I am typing this book must be connected to the main electricity supply. The amount of energy expended per bit of information depends on thermodynamic considerations. Dissipation is least when the processor operates at a temperature close to that of its environment. The human brain and most computers operate very inefficiently, and dissipate copious quantities of excess energy in the form of heat. The brain, for example, produces a sizable fraction of the body's heat, and many computers need a special cooling system to prevent them from melting. The origin of this waste heat can be traced to the very logic on which the information processing operates, which necessitates discarding information. For example, if a computer carries out the computation 1 + 2 = 3, then two bits of input information (1 and 2) are replaced by one bit of output information (3). Once the computation has been performed, the computer may discard the input information, thus replacing two bits by one. Indeed, to prevent its memory banks from clogging up, the machine has to discard such extraneous information all the time. The process of erasure is by definition irreversible, and therefore involves an increase in entropy. So it seems that on very basic grounds information gathering and processing will inevitably irreversibly deplete the available energy and raise the entropy of the universe.

Freeman Dyson has contemplated the limitations faced by a community of sentient beings—who are restricted by the need to dissipate energy at a certain rate, if only in order to think—as the universe cools toward a heat death. The first constraint is that the beings must have a temperature higher than that of their environment, otherwise the waste heat would not flow out of them. Secondly, the laws of physics limit the rate at which a physical system can radiate energy into its environment. Obviously, the beings cannot operate for long if they produce waste heat faster than they can get rid of it. These requirements place a lower limit on the rate at which the beings inevitably dissipate energy. An essential requirement is that there must exist a source of free energy to fuel this vital heat outflow. Dyson concludes that all such sources are destined to dwindle in the far cosmic future, so that all sentient beings eventually face an energy crisis.

Now, there are two ways of prolonging the longevity of sentience. One is to survive for as long as possible; the other is to speed up the *rate* of thinking and experiencing.

Dyson makes the reasonable assumption that a being's subjective experience of the passage of time depends on the rate at which the being processes information: the faster the processing mechanism used, the more thoughts and perceptions the being has per unit time, and the faster time appears to pass. This assumption is used in an entertaining manner in the science-fiction novel *Dragon's Egg,* by Robert Foreword, which tells the story of a community of conscious beings who live on the surface of a neutron star. These beings utilize nuclear rather than chemical processes to sustain their existence. Because nuclear interactions are thousands of times faster than chemical interactions, the neutronic beings process information much more rapidly. One second on the human time scale represents the equivalent of many years for them. The neutron-star community is fairly primitive when first contacted by humans but develops by the minute and soon overtakes humanity.

Unfortunately, adopting this strategy as a means of surviving in the far future has a downside: the faster the information is processed, the greater will be the rate of energy dissipation and the more rapidly will the available energy resources become depleted. You might think that this would inevitably spell doom for our descendants, whatever physical form they might adopt. But not necessarily so. Dyson has shown that there could be a clever compromise, in which the community gradually slows its rate of activity to match the running down of the universe—by, say, going into hibernation for ever-increasing lengths of time. During each somnolent phase, the heat from the endeavors of the previous active phase would be allowed to dissipate and useful energy to accumulate, for utilization in the next active phase.

The subjective time experienced by the beings who adopt this strategy will represent a smaller and smaller fraction of the actual time elapsed, because the downtime of the community is always getting longer. But, as I keep remarking, forever is a long time, and we have to contend with opposing limits: resources tending to zero and time tending to infinity. Dyson showed, from a simple examination of these limits, that the total subjective time can be infinite even if the total resources are finite. He quotes an amazing statistic: a community of beings with the same population level as humanity today could endure for literally eternity using a total energy of 6×10^{30} joules, this being the output from the sun for a period of only eight hours!

True immortality, however, demands more than the ability to process an infinite amount of information. If a being has a finite number of brain states, it can think only a finite number of different thoughts. If it were to endure forever, this would mean that the same thoughts would be entertained over and over again. Such an existence seems as pointless as that of a doomed species. To escape from this dead end, it is necessary for the community—or the single superbeing—to go on growing without limit. This poses a severe challenge in the very far future, because matter will be evaporating away faster than it can be commandeered as brain stuff. Perhaps a desperate but ingenious individual would attempt to harness the elusive but ever-present cosmic neutrinos to expand the scope of its intellectual activity.

Much of Dyson's discussion—and, indeed, most speculation about the fate of conscious beings in the far future—tacitly assumes that the mental processes of these beings always boil down to some sort of digital computing process. A digital computer is certainly a finite-state machine, and therefore faces a strict limit on what it can achieve. There are, however, other sorts of systems, known as analog computers. A simple example is a slide rule. Computations can be made by adjusting the rule continuously, and in an idealized case there can be an infinite number of states. Thus analog computers escape some of the limitations of digital computers, which can store and process only a finite amount of information. If information is encoded after the fashion of an analog computer—say, through the positions or angles of material objects—the capacity of the computer seems unlimited. So if a superbeing can operate as an analog computer, perhaps it can think not only an infinite number of thoughts but an infinite number of *different* thoughts.

Unfortunately, we do not know whether the universe as a whole is like an analog or a digital computer. Quantum physics suggests that the universe itself should be "quantized"—that is, discrete jumps rather than continuous variations are built into all its properties. But this is pure conjecture. Nor do we really understand the relationship between mental and physical brain activity; it may be that thoughts and experiences cannot be simply related to the quantum-physical ideas considered here.

Whatever the nature of mind may be, there is no doubt that the beings of the far future face the ultimate ecological crisis: the cosmic dissipation of all energy sources. Nevertheless, it appears that by "living it down" they could achieve a sort of immortality. In Dyson's scenario, their activities would impact less and less on a universe coldly indifferent to their requirements, and for untold eons they would rest inactive, retaining their memories but not adding to them, barely disturbing the still blackness of a moribund cosmos. By clever organization, they could still think an infinite number of thoughts and experience an infinite number of experiences. What more could we hope for?

The cosmic heat death has been one of the abiding myths of our age. We saw how Russell and others seized upon the seemingly inevitable degeneration of the universe as predicted by the second law of thermodynamics to support a philosophy of atheism, nihilism, and despair.

With our improved understanding of cosmology, we can today paint a somewhat different picture. The universe may be running down, but it is not running out. The second law of thermodynamics certainly applies, but it does not necessarily preclude cultural immortality.

In fact, things may not even be as bad as Dyson's scenario. So far I have assumed that the universe remains more or less uniform as it expands and cools, but this may not be correct. Gravitation is the source of many instabilities, and the large-scale uniformity of the cosmos we see today could give way to a more complicated arrangement in the far future. For example, slight variations in the rate of expansion in different directions could become amplified. Huge black holes might cluster together as their mutual attraction overcame the dispersing effect of the cosmological expansion. This circumstance would lead to a curious competition: Remember that the smaller a black hole is, the hotter it is and the faster it evaporates. If two black holes coalesce, the final hole will be larger, hence cooler, and the evaporation process will receive a major setback. The key question with regard to the far future of the universe is whether the rate of merging of black holes is sufficient to keep pace with the rate of evaporation. If it is, then there will always exist some black holes that can provide, by means of their Hawking radiation, a source of useful energy for a technologically adept community, possibly removing the need for hibernation. Calculations by the physicists Don Page and Randall McKee suggest that this competition is a knife-edge affair and depends critically on the precise rate at which the expansion of the universe continues to decline; in some models, black-hole coalescence does indeed win out.

Also neglected in Dyson's account is the possibility that our descendants may themselves attempt to modify the large-scale organization of the cosmos so as to preserve their longevity. The astrophysicists John Barrow and

Frank Tipler have considered ways in which an advanced technological community might make slight adjustments to the motions of stars in order to engineer a particular gravitational arrangement favorable to themselves. For example, nuclear weapons could be used to perturb the orbit of an asteroid—enough, say, so that it would receive a slingshot boost from a planet and go crashing into the sun. The momentum of that impact would very slightly alter the sun's orbit in the galaxy. Although the effect is small, it is cumulative: the farther the sun moves, the greater the displacement achieved. Over a distance of many light-years, the shift could make a crucial difference if the sun were to approach another star, changing a mere nodding encounter to one that violently modified the sun's trajectory across the galaxy. By manipulating many stars, clusters of astronomical bodies could be created and managed for the benefit of the community. And because the effects amplify and accumulate, there is no limit to the size of systems that can be controlled in this way—by a little nudge here and another there. Given long enough—and our descendants certainly have plenty of time at their disposal—even whole galaxies could be maneuvered.

This grandiose cosmic engineering would have to compete with the natural, random occurrences in which stars and galaxies are flung out of gravitationally bound clusters, as described in chapter 7. Barrow and Tipler find that it would take 10^{22} years to rearrange a galaxy by means of asteroid manipulation. Unfortunately, natural disruption occurs in about 10^{19} years, so the battle looks decidedly weighted in nature's favor. On the other hand, our descendants may gain control over much larger objects than asteroids. Also, the rate of natural dispersal depends on the orbital speeds of the objects. When it comes to whole galaxies, these speeds drop as the universe expands. The slower speeds also make artificial manipulation slower, but the two effects don't diminish at the same rate. It seems that, with time, the natural disruption rate might fall below the rate at which a community of engineers could reorder the universe. This raises the interesting possibility that as time goes on intelligent beings can gain more and more control over a less and less resourceful universe, until all of nature is essentially "technologized," and the distinction between what is natural and what is artificial disappears.

A key assumption of Dyson's analysis is that thought processes inevitably dissipate energy. Human thought processes definitely do, and until recently it was assumed that any form of information processing had to pay a minimum thermodynamic price. Surprisingly, that is not strictly correct. The computer scientists Charles Bennett and Rolf Landauer of IBM have demonstrated that reversible computation is possible in principle. This means that certain (at present entirely hypothetical) physical systems could process information without dissipation. It is possible to conceive of a system thinking an infinite number of thoughts without needing any sort of power supply! It is not clear that such a system could *gather* as well as process information, because any nontrivial information acquisition from the environment would seem to involve energy dissipation in one form or another, if only because it requires sorting out the signal from the noise. Therefore this undemanding being could have no perceptions of the world about it. It could, however, remember a universe that was. Perhaps it could also dream.

The image of the dying universe has obsessed scientists for over a century. The assumption that we are living in a cosmos steadily degenerating through entropic profligacy is part of the folklore of scientific culture. But how well established is it? Can we be sure that *all* physical processes inevitably lead toward chaos and decay?

What about biology? The extremely defensive manner with which some biologists defend Darwinian evolution gives a hint. Their reaction stems, I believe, from the uncomfortable contradiction of a process that is clearly constructive driven by physical forces that are supposed to be, at rock bottom, destructive. Life on Earth probably began as some sort of primeval slime. Today the biosphere is a rich and complex ecosystem, a network of elaborately complicated and highly varied organisms in subtle interaction. Although biologists, perhaps fearful of overtones of divine purpose, deny any evidence for systematic progress in evolution, it is clear to scientist and nonscientist alike that something has advanced, more or less unidirectionally, since life originated on Earth. The problem is to characterize that advance more sharply. What precisely is it that has advanced?

The foregoing discussions concerning survival have focused on the struggle between information (or order) and entropy—with entropy always gaining the upper hand. But is information *per se* the quantity we should be concerned about? After all, working your way systematically through all possible thoughts is about as thrilling as reading the telephone directory. What counts is surely the *quality* of experience—or, more generally, the quality of information that is gathered and utilized.

As far as we can tell, the universe started out in a more or less featureless state. With time, the richness and variety of physical systems we see today has emerged. The history of the universe is therefore the history of the growth of organized complexity. This seems like a paradox. I began my account by describing how the second law of thermodynamics tells us that the universe is dying, sliding inexorably from an initial state of low entropy to a final state of maximum entropy and zero prospects. So are things getting better or getting worse?

There is actually no paradox, because organized complexity is different from entropy. Entropy, or disorder, is the negative of information, or order: the more information you process—that is, the more order you generate—the greater the entropic price paid: order here gives rise to disorder somewhere else. Such is the second law; entropy always wins. But organization and complexity are not merely order and information. They refer to certain *types* of order and information. We recognize an important distinction between say, a bacterium and a crystal. Both are ordered, but in a different way. A crystal lattice represents regimented uniformity—starkly beautiful but essentially boring. By contrast, the elaborately arranged organization of a bacterium is richly interesting.

These seem like subjective judgments, but they can be stiffened with mathematics. In recent years, a whole new field of research has opened up which sets as its goal the quantification of such concepts as organized complexity, and seeks to establish general principles of organization to place alongside existing laws of physics. The subject is still in its infancy, but it is already challenging many traditional assumptions about order and chaos.

In my book *The Cosmic Blueprint,* I propose that a sort of "law of increasing complexity" operates in the universe, standing alongside the second law of thermodynamics. There is no incompatibility between these two laws. In practice, an increase in the organizational complexity of a physical system increases entropy. For example, in biological evolution a new, more complex organism emerges only after a lot of destructive physical and biological processes have occurred (the premature death of maladapted mutants, for example). Even the formation of a snowflake creates waste heat that drives up the entropy of the universe. But, as explained, the trade-off is not direct, because organization is not the negative of entropy.

I am greatly heartened to find that many other researchers have arrived at similar conclusions, and attempts are being made to formulate a "second law" of complexity. Although compatible with the second law of thermodynamics, the complexity law gives a very different account of cosmic change, describing a universe *progressing* (in some sense to be made rigorous by the investigations I alluded to) from largely featureless beginnings to ever more elaborately complex states.

In the context of the end of the universe, the existence of a law of increasing complexity has a profound significance. If organized complexity is not the opposite of entropy, then the limited store of negative entropy in the universe need not place a bound on the level of complexity. The entropic price paid for the advance of complexity may be purely incidental—rather than fundamental, as is the case with mere ordering or information processing. If this is so, then our descendants may be able to achieve states of ever greater organizational complexity without squandering dwindling resources. Though they may be restricted in the quantity of information they process, there may be no limit on the richness and quality of their mental and physical activities.

In this chapter and the last, I have tried to provide a glimpse of a universe slowing down but perhaps never quite running out of steam completely, of bizarre science-fiction creatures eking out an existence against odds that become stacked forever higher against them, testing their ingenuity against the inexorable logic of the second law of thermodynamics. The image of their desperate but not necessarily futile struggle for survival may exhilarate some readers and depress others. My own feelings are mixed.

The entire speculation is, however, predicated on the assumption that the universe will continue to expand forever. We have seen how this is only one possible fate for the cosmos. If the expansion decelerates fast enough, the universe may one day stop expanding and start to contract toward a big crunch. What hope for survival then?

CHAPTER 9

• •

LIFE IN THE FAST LANE

No amount of human or alien ingenuity can prolong life forever unless there *is* a "forever." If the universe can exist for a finite time only, then Armageddon is unavoidable. In chapter 6, I explained how the ultimate fate of the cosmos hinges on its total weight. Observations suggest that the weight of the universe lies very near the critical borderline between eternal expansion and eventual collapse. If the universe does eventually start to contract, the experiences of any sentient beings will be very different indeed from the description given in the last chapter.

The early stages of cosmological contraction are not in the least threatening. Like a ball reaching the top of its trajectory, the universe will start its inward fall very slowly. Let us suppose for the moment that the high point is reached in a hundred billion years' time: there will still be plenty of stars burning then, and our descendants will be able to follow the motions of galaxies with optical telescopes—watching as the galactic clusters gradually slow in their retreat and then begin falling back toward each other. The galaxies we see today will be about four times farther away at that time. Because of the greater age of the universe, astronomers will be able to see about ten times as far as we can, so their observable universe will encompass many more galaxies than are visible to us at our cosmic epoch.

The fact that light takes many billions of years to traverse the cosmos means that any astronomers a hundred billion years hence will not see the contraction for a very long time. They will first notice that relatively nearby galaxies are, on average, more often approaching than receding, but the light from distant galaxies will still appear to be redshifted. Only after tens of billions of years would a systematic inrush become apparent. More easily recognizable would be a subtle change in the temperature of the cosmic background heat radiation. Recall that this background radiation is left over from the big bang and currently has a temperature of about three degrees above absolute zero, or 3°K. It cools as the universe expands. In a hundred billion years, it will have fallen to about 1°K. The temperature will bottom out at the high point of the expansion, and as soon as the contraction sets in it will begin to rise again, returning to 3°K when the universe has contracted to the density it has today. This will take another hundred billion years: the rise and fall of the universe is approximately symmetric in time.

The universe doesn't simply collapse overnight. In fact, for tens of billions of years our descendants will be able to make a good living, even after the contraction has begun. The situation is not quite so rosy, however, if the turnaround occurs after a much longer duration—say, a trillion trillion years. In this case, the stars will have burned out before the high point is reached, and any surviving inhabitants will be facing many of the same problems encountered in an ever-expanding universe.

Whenever the turnaround occurs, as measured in years from now, after the same number of years again the universe will have returned to its present size. Its appearance will be very different, though. Even with the turnaround at a hundred billion years, there will be many more black holes and many fewer stars than there are today. Habitable planets will be at a premium.

By the time the universe returns to its present size, it will be contracting at quite a pace, halving its dimensions in about three and a half billion years and accelerating all the time. The fun really starts about ten billion years after this point, however, when the rise in temperature of the cosmic background heat radiation will have become a serious threat. By the time the temperature has risen to about 300°K, a planet like the Earth would have difficulty divesting itself of heat. It would begin relentlessly warming up. First, any ice caps or glaciers would melt, then the oceans would start to evaporate.

Forty million years later, the temperature of the background radiation would reach the average temperature of Earth today. Earthlike planets would then be completely inhospitable. Of course, the Earth will have already faced such a fate, because the sun will have expanded to become a red giant, but now there is nowhere else for our descendants to go, no safe haven. The heat radiation fills the universe. All of space has a temperature of 300°C and rising. Any astronomers who had adapted to the torrid conditions, or created refrigerated ecosystems to delay being cooked, would notice that the universe was now collapsing at a hectic pace, halving in size every few million years. Any galaxies that still existed would no longer be recognizable, because they would by now have merged. However, there would still be a lot of empty space: collisions between individual stars would be rare.

The conditions in the universe as it approached its final phase would increasingly resemble those that prevailed shortly after the big bang. The astronomer Martin Rees has carried out an eschatological study of the collapsing cosmos. By applying general physical principles, he has been able to build up a picture of the final stages of collapse. Eventually, the cosmic heat radiation would become so intense that the night sky would glow dull red. The universe would slowly transform itself into an all-encompassing cosmic furnace, grilling all fragile life-forms wherever they might be hiding, and stripping away planetary atmospheres. Gradually, the red glow would turn yellow and then white, until the fierce heat radiation bathing the universe would threaten the existence of the stars themselves. Unable to radiate away their energy, the stars would build up heat inside and explode. Space would become filled with hot gas—plasma—glowing fiercely and getting hotter all the time.

As the pace of change quickens, so conditions become ever more extreme. The universe begins to change appreciably on a time scale of only a hundred thousand years, then a thousand, then a hundred, accelerating toward total catastrophe. The temperature rises to millions, then billions of degrees. Matter that occupies vast regions of space today is squeezed into tiny volumes. The mass of a galaxy occupies a space just a few light-years across. The last three minutes have arrived.

The temperature eventually becomes so great that even atomic nuclei disintegrate. Matter is stripped down to a uniform soup of elementary particles. The handiwork of the big bang, and of generations of stars in creating heavy chemical elements, is undone in less time than it takes you to read this page. Atomic nuclei—stable structures that may have endured for trillions of years—are irreversibly smashed. With the exception of black holes, all other structures have long since been seared into nonexistence. The universe now has an elegant but sinister simplicity. It has but seconds to live.

As the cosmos collapses faster and faster, the temperature rises without known limit at an escalating rate. Matter is compressed so strongly that individual protons and neutrons no longer exist; there is only a soup of quarks. Still the collapse accelerates.

The stage is now set for the ultimate cosmic catastrophe, a few microseconds away. Black holes begin to merge with each other, their interiors little different from the general collapsing state of the universe itself. They are now merely spacetime regions that have reached the end a little early and are being joined by the rest of the cosmos.

In the final moments, gravity becomes the all-dominant force, mercilessly crushing matter and space. The curvature of spacetime increases ever faster. Larger and larger regions of space are compressed into smaller and smaller volumes. According to conventional theory, the implosion becomes infinitely powerful, crushing all matter out of existence and obliterating every physical thing, including space and time themselves, at a spacetime singularity.

This is the end.

The "big crunch," as far as we understand it, is not just the end of matter. It is the end of *everything.* Because time itself ceases at the big crunch, it is meaningless to ask what happens next, just as it is meaningless to ask what happened before the big bang. There is no "next" for anything at all to happen—no time even for inactivity nor space for emptiness. A universe that came from nothing in the big bang will disappear into nothing at the big crunch, its glorious few zillion years of existence not even a memory.

Should we be depressed by such a prospect? Which is worse: a universe slowly degenerating and expanding forever toward a state of dark emptiness, or one that implodes to fiery oblivion? And what hope for immortality now, in a universe destined to run out of time?

Life in the approach to the big crunch looks even more hopeless than in the far future of an ever-expanding universe. The problem now is not a lack of energy but an excess of it. However, there may be billions or even trillions of years for our descendants to prepare for the final holocaust. During this time, life could expand throughout the cosmos. In the simplest model of a collapsing universe, the total volume of space is actually finite. This comes about because space is curved and can connect with itself in the three-dimensional equivalent of the surface of a sphere. It is therefore conceivable that intelligent beings can spread throughout the universe and gain control of it, thereby positioning themselves to confront the big crunch with all possible resources at their disposal.

At first, it is hard to see why they should bother. Given that existence beyond the big crunch is an impossibility, what would be the point in prolonging the agony just a little bit longer? Annihilation ten million or one million years before the end is all the same in a universe trillions of years old. But we must not forget that time is relative. The subjective time of our descendants will depend on their rate of metabolism and information processing. Again, assuming that they have plenty of time to adapt their physical form, they may be able to turn the approach of Hades into a type of immortality.

A rising temperature means that particles move more quickly and physical processes happen faster. Remember that the essential requirement of a sentient being is the ability to process information. In a universe with escalating temperature, the information-processing rate will also accelerate. To a being utilizing thermodynamic processes at one billion degrees, the imminent obliteration of the universe will appear to be years away. No need to be afraid of the end of time if the remaining time can be infinitely stretched in the minds of the observers. As the collapse accelerates toward the final crunch, so the subjective experiences of observers could in principle dilate ever faster, matching the accelerating plunge to Armageddon with an escalating speed of thought. Given sufficient resources, these beings would literally be able to buy time.

One might wonder whether a superbeing inhabiting the collapsing universe in its final moments could have an infinite number of distinct thoughts and experiences in the finite time available. This question has been studied by John Barrow and Frank Tipler. The answer depends critically on the physical details of the final stages. For example, if the universe remains fairly uniform in its approach to the final singularity, a major problem arises. Whatever the speed of thought, the speed of light remains unchanged, and light can travel at most a distance of one light-second per second. Because the speed of light defines the limiting speed at which *any* physical effect may propagate, it follows that no communication can take place between regions of the universe more than one light-second apart during the final second. (This is another example of an event horizon, similar to the one that prevents information getting out of black holes.) As the end approaches, so the size of communicable regions and the numbers of particles they contain shrink toward zero. For a system to process information, all parts of the system need to communicate. Clearly, the finite speed of light acts to restrict the size of any "brain" that may exist as the end approaches, and this in turn could limit the number of distinct states—hence thoughts—such a brain may have.

To evade this restriction, it is necessary for the final stages of cosmic collapse to deviate from uniformity—and in fact this eventuality is very probable. Extensive mathematical investigations of gravitational collapse suggest that as the universe implodes, the rate of collapse will vary in different directions. Curiously, it is not simply a matter of the universe shrinking faster in one direction than another. What happens is that oscillations set in, so that the direction of most rapid collapse keeps changing. In effect, the universe wobbles its way toward extinction in cycles of ever-increasing violence and complexity.

Barrow and Tipler conjecture that these complicated oscillations cause the event horizon to disappear first in this direction then in that, enabling all regions of space to keep in touch. Any superbrain would need to be quick-witted and switch communications from one direction to another as the oscillations brought more rapid collapse in one direction and then another. If the being can keep pace, the oscillations could themselves provide the necessary energy to drive the thought processes. Furthermore, in simple mathematical models there appears to be an infinite number of oscillations in the finite duration terminating in the big crunch. This provides for an infinite amount of information processing, hence, by hypothesis, an infinite subjective time for the superbeing. Thus the mental world may never end, even though the physical world comes to an abrupt cessation at the big crunch.

What might a brain of unlimited capability do? According to Tipler, it would not only be able to deliberate on all aspects of its own existence and that of the universe it had engulfed but, with its infinite information-processing power, it could go on to simulate imaginary worlds in an orgy of virtual reality. There would be no limit to the number of possible universes it could internalize in this way. Not only would the last three minutes stretch to eternity but they would also permit the simulated reality of an infinite variety of cosmic activity.

Unfortunately, these (somewhat wild) speculations depend on very particular physical models, which may turn out to be totally unrealistic. They also ignore the quantum effects that would probably dominate the final stages of gravitational collapse—effects that might well set an ultimate limit to the rate of information processing. If so, let us hope that the cosmic superbeing or supercomputer will at least come to understand existence well enough in the available time to become reconciled to its own mortality.

··

CHAPTER 10

SUDDEN DEATH—AND REBIRTH

So far I have assumed that the end of the universe, whether by bang or whimper (or, more accurately, crunch or deep freeze), is set in the very distant, possibly the infinite, future. If the universe collapses, our descendants would have many billions of years' warning of the impending crunch. But there remains another, altogether more alarming possibility.

As I have explained, when astronomers peer at the heavens, they do not see the universe in its present state, displayed like an instantaneous snapshot. Because of the time that light takes to reach us from distant regions, we see any given object in space as it was when the light was emitted. The telescope is also a timescope. The farther away the object is situated, the farther back in time will be the image we see today. In effect, the astronomer's universe is a backward slice through space and time, known technically as the "past light cone," and depicted in figure 10.1.

According to the theory of relativity, no information or physical influence can travel faster than light. Therefore, the past light cone marks the limit not only of all knowledge about the universe but of all events that can possibly affect us at this moment. It follows that any physical influ-

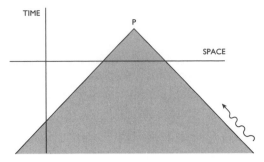

FIGURE 10.1

From a particular point *P* in space and time—which might be *here* and *now,* for example—an astronomer looking out into the universe actually sees the universe as it was in the past, not as it is now. The information arriving at *P* travels up along the "past light cone" through *P,* marked by the oblique lines. These are the paths of light signals converging on Earth from distant regions of the universe in the past. Because no information or physical influence can travel faster than light, the observer at the moment depicted can know only about influences or events happening in the shaded region. An apocalyptic event outside the past light cone might be sending disastrous influences (wavy line) racing toward Earth, but the observer would be blissfully unaware of this until the influences arrived.

• • • • • • • • •

ence coming at us at the speed of light comes entirely without warning. If catastrophe is heading our way up the past light cone, there will be no harbinger of doom. The first we will know about it will be when it hits us.

To give a simple hypothetical example, if the sun were to blow up now, we would not be aware of the fact until about eight and a half minutes later, this being the time it takes for light to reach us from the sun. Similarly, it is entirely possible that a nearby star has already blown up as a supernova—an event that might bathe Earth in deadly radiation—but that we shall remain in blissful ignorance of the fact for a few more years yet while the bad news races across the galaxy at the speed of light. So although the universe may look quiet enough at the moment, we can't be sure that something really horrible hasn't already happened.

Most sudden violence in the universe entails damage that is limited to the immediate cosmic locality. The death of stars or the plunge of matter into a black hole will disrupt planets and nearby stars, perhaps as far as a few light-years away. The most spectacular outbursts seem to be events that befall the cores of some galaxies. As I have described, huge jets of material are sometimes ejected at a large fraction of the speed of light, and prodigious quantities of radiation are also emitted. This is violence on a galactic scale.

But what about events of universe-wrecking proportions? Is it possible that a convulsion can occur that would destroy the entire cosmos at a stroke—in midlife, so to speak? Could a truly cosmic catastrophe already have been triggered, its unpleasant effects even now sweeping up our past light cone toward our fragile niche in space and time?

In 1980, the physicists Sidney Coleman and Frank De Luccia published a portentous paper with the innocuous title "Gravitational Effects on and of Vacuum Decay" in the journal *Physical Review D*. The vacuum to which they refer is not merely empty space but the vacuum state of quantum physics. In chapter 3, I explained that what appears to us as emptiness is in reality seething with ephemeral quantum activity, as ghostly virtual particles appear and disappear again in a random frolic. Recall that this vacuum state may not be unique; there could be several quantum states, all appearing empty but enjoying different levels of quantum activity and different associated energies.

It is a well-established principle of quantum physics that higher-energy states tend to decay into lower-energy states.

An atom, for example, may exist in a range of excited states, all of which are unstable, and will try to decay to the lowest energy, or "ground," state, which is stable. Similarly, an excited vacuum will try to decay to the lowest energy, or "true," vacuum. The inflationary-universe scenario is based on the theory that the very early universe had an excited, or "false," vacuum state, during which time it inflated frenetically, but that in a very short time this state decayed to the true vacuum and inflation ceased.

The usual assumption is that the present state of the universe corresponds to the true vacuum; that is, empty space at our epoch is the vacuum with the lowest possible energy. But can we be sure of that? Coleman and De Luccia consider the chilling possibility that the present vacuum may be not the true vacuum but merely a long-lived, metastable, false vacuum that has lulled us into a false sense of security because it has endured for a few billion years. We know of many quantum systems, such as uranium nuclei, that have half-lives of billions of years. Suppose the present vacuum falls into this category? The "decay" of the vacuum mentioned in the title of Coleman and De Luccia's paper refers to the catastrophic possibility that the present vacuum may suddenly fail and pitch the cosmos into an even lower energy state, with dire consequences for us (and all else besides).

The key to the Coleman and De Luccia hypothesis is the phenomenon of quantum tunneling. This can best be illustrated with the simple case of a quantum particle trapped by a barrier of force. Suppose the particle sits in a little valley bounded on either side by hills, as shown in figure 10.2. Of course, these don't have to be real hills; they could be electric or nuclear force fields, for example. In the absence of the energy needed to surmount the hills (or overcome the force barrier), the particle appears to be trapped forever. But recall that all quantum particles are subject to Heisenberg's uncertainty principle, which per-

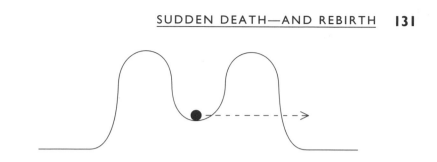

FIGURE 10.2

Tunnel effect. If a quantum particle is trapped in a valley between two hills, there is a small probability that it can escape by borrowing energy and hopping over the hill. In effect, it is observed to tunnel through the barrier. A familiar case occurs when alpha particles in the nuclei of certain elements tunnel through the nuclear force barrier and fly away, a phenomenon known as alpha radioactivity. In this example, the "hill" is due to nuclear and electric forces, and the picture drawn here is schematic only.

• • • • • • • • •

mits energy to be borrowed for small durations. This opens up an intriguing possibility. If the particle can borrow enough energy to reach the top of the hill and get across to the other side before having to pay the energy back, it can escape from the well. In effect, it will have "tunneled" through the barrier.

The probability for a quantum particle to tunnel out of a well like this depends very sensitively on both the height and the width of the barrier. The higher the barrier, the more energy the particle must borrow to reach the top, and so, according to the uncertainty principle, the shorter the duration of the loan must be. Hence high barriers can be tunneled through only if they are also thin, enabling the particle to traverse them quickly enough to repay the loan on time. For this reason, the tunnel effect is not noticed in daily life: macroscopic barriers are far too high and wide for significant tunneling to occur. In principle, a human being can walk through a brick wall, but the quantum-tunneling probability for this miracle is exceedingly small. On an atomic scale, however, tunneling is very common; for example, it is the mechanism by which alpha radioactivity occurs. The tunnel effect is also exploited in semiconductors and other electronic devices, such as the scanning tunneling electron microscope.

With regard to the problem of the possible decay of the present vacuum, Coleman and De Luccia speculate that the quantum fields making up the vacuum might be subject to a (metaphorical) landscape of forces like that shown in figure 10.3. The present vacuum state corresponds to the base of valley A. The true vacuum, however, corresponds to the base of valley B, which is lower than A. The vacuum would like to decay from the higher energy state A into the lower energy state B, but it is deterred

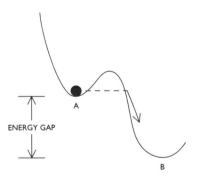

FIGURE 10.3

False and true vacuum states. It may be the case that the present quantum state of empty space A is not the lowest energy state, but that it is nevertheless quasi-stabilized by corresponding to a sort of high-altitude valley. There would then be a small probability of the state decaying by the tunnel effect to the truly stable ground state B. The transition between these states, occurring via bubble formation, would release a vast quantity of energy.

from so doing by the "hill," or force field, that separates them. Although the hill impedes decay, it does not entirely prevent it, on account of the tunnel effect: the system can tunnel through from valley A to valley B. If this theory is correct, then the universe is living on borrowed time, hung up in valley A, but with an ever-present chance that it will tunnel into valley B at some arbitrary moment.

Coleman and De Luccia were able to model the decay of the vacuum mathematically—to trace the manner in which the phenomenon occurs. They found that decay will start at a random location in space, in the form of a tiny bubble of true vacuum surrounded by the unstable false vacuum. As soon as the bubble of true vacuum has formed, it will expand at a rate that rapidly approaches the speed of light, engulfing a larger and larger region of the false vacuum and instantaneously converting it into true vacuum. The energy difference between the two states—which might have the sort of enormous value I discussed in chapter 3—is concentrated in the bubble wall, which sweeps across the universe spelling destruction to everything in its path.

The first we would know about the existence of a true-vacuum bubble would be when the wall arrived and the quantum structure of our world suddenly changed. We wouldn't even have three minutes' warning. Instantaneously, the nature of all subatomic particles and their interactions would alter drastically; for example, protons might immediately decay, in which case all matter would abruptly evaporate. What was left would then find itself inside the bubble of true vacuum—a state of affairs very different from what we observe at the moment. The most significant difference concerns gravitation. Coleman and De Luccia found that the energy and pressure of the true vacuum would create a gravitational field so intense that the region embraced by the bubble would collapse, even as the bubble wall expands, in less than microseconds. No gentle fall toward a big crunch this time; instead, abrupt annihilation of everything, as the bubble interior implodes into a spacetime singularity. In short, instant crunch. "This is disheartening," remark the authors, in a masterful understatement, and they continue:

> The possibility that we are living in a false vacuum has never been a cheering one to contemplate. Vacuum decay is the ultimate ecological catastrophe; . . . after vacuum decay not only is life as we know it impossible, so is chemistry as we know it. However, one could always draw stoic comfort from the possibility that perhaps in the course of time the new vacuum would sustain, if not life as we know it, then at least some structures capable of knowing joy. This possibility has now been eliminated.

The appalling consequences of vacuum decay became the subject of much discussion among physicists and astronomers following the publication of Coleman and De Luccia's paper. In a follow-up study published in the journal *Nature*, the cosmologist Michael Turner and the physicist Frank Wilczek arrived at an apocalyptic conclusion: "From the point of view of microphysics, then, it is quite conceivable that our vacuum is metastable . . . without warning a bubble of true vacuum could nucleate somewhere in the Universe and move outwards at the speed of light."

Shortly after the Turner and Wilczek paper appeared, Piet Hut and Martin Rees, also writing in *Nature*, raised the alarming specter that the formation of a universe-destroying vacuum bubble might be inadvertently triggered by particle physicists themselves! The worry is that the very high-energy collision of subatomic particles might create conditions—just for an instant, in a very small region of space—which would encourage the vacuum to decay. Once the transition had occurred, even on a microscopic scale, there would be no stopping the newly formed bubble from rapidly ballooning to astronomical proportions. Should we place a ban on the next generation of particle accelerators? Hut and Rees gave welcome reassurance, pointing out that cosmic rays achieve higher energies than we can make inside our particle accelerators, and that these cosmic rays have been hitting nuclei in the Earth's atmosphere for billions of years without triggering vacuum decay. On the other hand, with an improvement by a factor of a few hundred or so in accelerator energies, we might be capable of creating collisions more energetic than any that have occurred from cosmic-ray impacts on Earth. The real issue, however, is not whether bubble formation could occur on Earth but whether it has occurred anywhere in the observable universe at any time since the big bang. Hut and Rees noted that on very rare occasions two cosmic rays will suffer a head-on collision, with energies a billion times higher than those possible in existing accelerators. So we don't need a regulatory authority yet.

Paradoxically, vacuum-bubble formation—the same phenomenon that threatens the very existence of the cosmos—could, in a slightly different context, prove to be its inhabitants' only feasible salvation. The one sure way to escape the death of the universe is to create a new one and escape into it. This may sound like the last word in fanciful speculation, but "baby universes" have been much discussed in recent years, and the argument for their existence has a serious side to it.

The subject was originally raised by a group of Japanese physicists in 1981, who studied a simple mathematical model of the behavior of a small bubble of false vacuum surrounded by true vacuum—a situation the inverse of that just discussed. What was predicted is that the false vacuum would inflate in the manner described in chapter 3, rapidly expanding into a large universe in a big bang. At first, it seems that the inflation of the false-vacuum bubble must cause the bubble wall to expand so that the region of false vacuum grows at the expense of the region of true vacuum. But this contradicts the expectation that it is the lower energy true vacuum that should displace the higher energy false vacuum and not the other way about.

Oddly, viewed from the true vacuum, the region of space occupied by the bubble of false vacuum does not appear to inflate. In fact, it looks more like a black hole. (In this it resembles the Tardis, Dr. Who's time machine, which appears bigger on the inside than it does on the outside.) A hypothetical observer situated inside the false-vacuum bubble would see the universe swell to enormous proportions, but, viewed from outside, the bubble remains compact.

One way to envisage this peculiar state of affairs is by analogy with a rubber sheet that blisters up in one place and balloons out (see figure 10.4). The balloon forms a sort of baby universe connected to the mother universe by an umbilical cord, or "wormhole." The throat of the wormhole appears, from the mother universe, as a black hole. This configuration is unstable; the black hole quickly evaporates by the Hawking effect, and disappears from the mother universe completely. As a result, the wormhole is pinched off, and the baby universe, now disconnected from the mother universe, becomes a new and independent universe in its own right. The development of the child universe following this budding-off from the mother is the same as it supposedly was for our universe: a brief period of inflation followed by the usual deceleration. The model carries the obvious implication that our own universe may have originated in this way—as the progeny of another universe.

Alan Guth, the originator of the inflationary theory, and his colleagues have investigated whether the preceding scenario permits the bizarre possibility of creating a new universe in the laboratory. Unlike the scary case of the

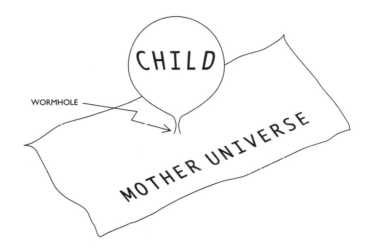

FIGURE 10.4

A bubble of space balloons out from the mother universe to form a child universe, connected to the mother by an umbilical wormhole. From the viewpoint of the mother universe, the mouth of the wormhole would appear to be a black hole. As the black hole evaporates, the throat of the wormhole pinches off, disconnecting the baby universe, which then leads an independent existence as a universe in its own right.

· · · · · · · · ·

decay of a false vacuum into a bubble of true vacuum, the creation of a bubble of false vacuum surrounded by true vacuum does not threaten the existence of the universe. Indeed, although the experiment may trigger a big bang, the explosion would be completely confined inside a tiny black hole, which soon evaporates. The new universe would create its *own* space, not eat up any of ours.

Although the idea remains highly conjectural and is based entirely on mathematical theorizing, some studies suggest that the creation of new universes in this way may be possible by concentrating large amounts of energy in a carefully sculpted manner. In the very far future, when our own universe is becoming uninhabitable or approaching a big crunch, our descendants may decide to get out for good by initiating the budding process and then scrambling through the umbilical wormhole into the universe next door before it pinches off—the ultimate in emigration. Of course, nobody has any idea how or whether these intrepid beings could accomplish this feat. At the very least, the journey through the wormhole would be pretty uncomfortable, unless the black hole into which they had to plunge was very large.

Ignoring such practical matters, the very possibility of baby universes opens up the prospect of genuine immortality—not just for our descendants but for universes too. Rather than thinking about the life and death of *the* universe, we ought instead to think about a family of universes multiplying ad infinitum, each giving birth to new generations of universes, perhaps in their legions. With such cosmic fecundity, the assemblage of universes—or the metaverse, as it should really be called—might have no beginning or end. Each individual universe would have a birth, evolution, and death in the manner described in the earlier chapters of this book, but the collection as a whole would exist eternally.

This scenario raises the question of whether the creation of our own universe was a natural affair (analogous to natural childbirth) or the result of deliberate manipulation (a "test-tube baby"). We can imagine that a sufficiently advanced and altruistic community of beings in a mother universe might decide to create baby universes not to provide an escape route for their own survival but merely to perpetuate the possibility of life existing somewhere, given that their own universe is doomed. This removes the need to tackle the formidable obstacles facing any attempt to construct a traversible wormhole into a child universe.

It is not clear to what extent the baby universe would carry the genetic imprint of its mother. Physicists do not yet have an understanding of why the various forces of nature and particles of matter have the properties they do. On the one hand, these properties might be part of the laws of nature, fixed once and for all in any universe. On the other hand, some of the properties may be the result of accidents of evolution. For example, there may well be several true-vacuum states, all with identical, or nearly identical, energy. It could be that when the false vacuum decays at the end of the inflationary era, it simply picks at random one of these many possible vacuum states. As far as the physics of the universe is concerned, the choice of vacuum state will dictate many of the properties of the particles and the forces that act between them, and could even dictate the number of spatial dimensions. So a baby universe might have completely different properties from its mother. Perhaps life will be possible only in a very small number of progeny, where the physics resembles that of our universe rather closely. Or perhaps there is a sort of heredity principle that insures that baby universes closely inherit the properties of their mother universes, save for the odd mutation. The physicist Lee Smolin has suggested that there could even be a type of Darwinian evolution operating among universes which indirectly encourages the emergence of life and consciousness. Even more interesting is the possibility that universes are created by intelligent manipulation in a mother universe and deliberately endowed with the necessary properties to give rise to life and consciousness.

None of these ideas amounts to much more than wild conjecture, but the subject of cosmology is still a very young science. The fanciful speculations considered above at least serve as an antidote to the gloomy prognoses developed in the earlier chapters. They hint at the possibility that even if our descendants must one day face the last three minutes, conscious beings of some sort may always exist somewhere.

··

CHAPTER 11

The bizarre ideas discussed at the end of the last chapter are not the only possibilities that have been mooted in the search for a way of avoiding cosmic doom. Whenever I give a lecture about the end of the universe, someone usually asks me about the cyclic model. The idea is this. The universe expands to a maximum size, then contracts to a big crunch, but instead of obliterating itself completely it somehow "bounces" and embarks on another cycle of expansion and contraction (see figure 11.1). This process might go on forever, in which case the universe would have no real beginning or end, even though each individual cycle would be marked by a distinct start and finish. It is a theory that particularly appeals to people who have been influenced by Hindu and Buddhist mythology, in which cycles of birth and death, creation and destruction, figure prominently.

I have outlined two very different scientific scenarios for the end of the universe. Each is disturbing in its own way. The prospect of the cosmos obliterating itself entirely in a big crunch is alarming, however far in the future this event may lie. On the other hand, a universe that lasts for an infinite time in a state of bleak emptiness after a finite duration of glorious activity is profoundly depressing. The

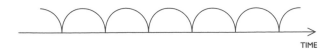

TIME

FIGURE 11.1

The cyclic-universe model. The universe pulsates in size in a periodic manner between very dense states and very distended states. Each cycle begins with a big bang and ends with a big crunch, and is approximately symmetric in time.

·········

fact that each model may possibly provide for superbeings to achieve unlimited information-processing power may seem cold comfort for us warm-blooded *Homo sapiens*.

The appeal of the cyclic model is that it evades the specter of total annihilation, without replacing it by eternal degeneration and decay. To avoid the futility of endless repetition, the cycles should be somehow different from each other. In one popular version of the theory, each new cycle emerges phoenixlike from the fiery death of its predecessor. From this pristine condition, it develops new systems and structures and explores its own rich novelty before the slate is wiped clean once more at the next big crunch.

Attractive though the theory may seem, it unfortunately suffers from grave physical problems. One of these is identifying a plausible process that will allow the collapsing universe to bounce at some very high density rather than to annihilate itself in a big crunch. There has to be some sort of antigravitational force that becomes overwhelmingly large at the late stages of collapse in order to reverse the momentum of implosion and counter the formidable crushing power of gravity. No such force is known at present, and if it existed its properties would have to be very strange.

The reader may recall that precisely such a powerful repulsive force is hypothesized in the inflationary theory of the big bang. However, remember that the excited vacuum state that produces the inflationary force is highly unstable, and soon decays. Although it is conceivable that the tiny, simple, nascent universe should have originated in such an unstable state, it is quite another thing to postulate that a universe shrinking from a complicated macroscopic condition could contrive to recover the excited vacuum state everywhere. The situation is analogous to balancing a pencil on its point. The pencil soon topples over; that is easy. Much harder would be to knock the pencil back onto its point once more.

Even supposing that such problems could be circumvented somehow, there remain serious difficulties with the cyclic-universe idea. One of these I discussed in chapter 2. Systems subject to irreversible processes that proceed at a finite rate will tend to approach their final state after a finite period of time. It was this principle that led to the prediction of universal heat death in the nineteenth century. Introducing cosmic cycles does not circumvent the difficulty. The universe can be compared to a clock, slowly running down. Its activity will inevitably eventually cease unless it somehow gets rewound. But what mechanism could rewind the cosmic clock without itself being subject to irreversible change?

At first sight, the collapse phase of the universe looks like a reversal of the physical processes that occur in the expanding phase. The dispersing galaxies are pulled back together, the cooling background radiation is reheated, and the complex elements are broken down to a soup of elementary particles again. The state of the universe just before the big crunch bears a strong similarity to its state just after the big bang. However, the impression of symmetry is only superficial. We gain a clue from the fact that astronomers living at the time of reversal, when expansion turns into contraction, go on seeing the distant galaxies recede for many billions of years. The universe *looks* as if it is still expanding, even though it is contracting. The illusion is due to the lag in appearances occasioned by the finite speed of light.

In the 1930s, the cosmologist Richard Tolman showed how this lag destroys the apparent symmetry of the cyclic universe. The reason is simple. The universe starts out with a lot of heat radiation left over from the big bang. Over time, starlight augments this radiation, so that after a few billion years there is almost as much energy in accumulated starlight pervading space as there is in the background heat. This means that the universe approaches the big crunch with considerably more radiation energy spread throughout it than was the case just after the big bang, so when the universe eventually contracts back to the same density that it has today, it will be somewhat hotter.

The extra heat energy is paid for by the matter content of the universe, through Einstein's $E = mc^2$ formula. Inside the stars that produce the heat energy, light elements such as hydrogen get processed into heavy elements such as iron. A nucleus of iron normally contains twenty-six protons and thirty neutrons. You might suppose that such a nucleus should therefore have the mass of twenty-six protons and thirty neutrons, but it doesn't. The assembled nucleus is about 1 percent lighter than the sum of the masses of the individual particles. The "missing" mass is accounted for by the large binding energy produced by the strong nuclear force; the mass represented by this energy is released to pay for starlight.

The upshot of all this is a net transfer of energy from matter to radiation. This has an important effect on the way the universe contracts, because the gravitational pull of radiation is quite different from that of matter of the same mass energy. Tolman showed that the extra radiation in the contracting phase causes the universe to collapse at a faster rate. If by some means a bounce were to occur, the universe would then emerge expanding at a faster rate too. In other words, each big bang would be bigger than the last. As a result, the universe would expand to a greater size with each new cycle, so the cycles would gradually get both bigger and longer. (See figure 11.2).

The irreversible growth of the cosmic cycles is no mystery. It is an example of the inescapable consequences of the second law of thermodynamics. The accumulating radiation represents a growth of entropy, which manifests itself gravitationally in the form of bigger and bigger cycles. It does, however, put an end to the idea of true cyclicity: the universe clearly evolves over time. Toward the past, the cycles cascade together into a complicated and messy beginning, while the future cycles expand without limit, until they become so long that any given cycle would be for the most part indistinguishable from the heat-death scenario of the ever-expanding models.

Since the work of Tolman, cosmologists have been able to identify other physical processes that break the symmetry of the expanding and contracting phases of each cycle. One example is the formation of black holes. In the standard picture, the universe begins without any black holes, but as time goes on stars collapse and other processes

FIGURE 11.2

Irreversible processes cause the cosmological cycles to grow and grow, thus destroying true cyclicity.

cause black holes to form. As the galaxies evolve, more and more black holes appear. During the late stages of the collapse, the compression will encourage the formation of yet more holes. Some of the black holes may merge to form larger holes. The gravitational arrangement of the universe near the big crunch is therefore much more complicated—indeed, distinctly more holey—than it was near the big bang. If the universe were to bounce, the next cycle would begin with many more black holes than this one.

The conclusion seems inescapable that any cyclic universe that allows physical structures and systems to propagate from one cycle to the next will not evade the degenerative influences of the second law of thermodynamics. There will still be a heat death. One way to sidestep this dismal conclusion is to suppose that the physical conditions at the bounce are so extreme that no information about earlier cycles can get through to the next. All preceding physical objects are destroyed, all influences annihilated. In effect, the universe is reborn entirely from scratch.

It is hard to see, however, what attraction such a model holds. If each cycle is physically disconnected from the others, what meaning does it have to say that the cycles succeed each other, or represent the *same* universe somehow enduring? The cycles are effectively distinct separate universes, and might just as well be said to exist in parallel rather than in sequence. The situation is reminiscent of the doctrine of reincarnation, whereby the reborn person has no memory of previous lives. In what sense can one say that the *same* person is reincarnated?

Another possibility is that the second law of thermodynamics is somehow violated, so that "the clock gets rewound" at the bounce. What does it mean for the damage caused by the second law to be undone? Let's take a simple example of the second law at work: the evaporation of perfume from a bottle, say. A reversal of fortunes for the perfume would entail a gigantic conspiracy of organization, in which every perfume molecule throughout the room was knocked back into the bottle. The "movie" would be played in reverse. It is from the second law of thermodynamics that we obtain the distinction between past and future—the arrow of time. A violation of the second law therefore amounts to a reversal of time.

It is, of course, a somewhat trivial evasion of cosmic death to suppose that time simply reverses when the crack of doom is heard. When the going gets rough, just run the great cosmic movie backward! Nevertheless, the idea has appealed to some cosmologists. In the 1960s, the astrophysicist Thomas Gold suggested that time might run backward in the contracting phase of a recontracting universe. He pointed out that such a reversal would include the brain functions of any beings around at the time, and so serve to reverse their subjective sense of time. The inhabitants of the contracting phase would not, therefore, see everything around them "running backward" but would experience the forward flow of events in the same manner as we do. For example, they would perceive the universe to be expanding, not contracting. Through their eyes, it would be *our* phase of the universe that was contracting and our brain processes that were running backward.

In the 1980s, Stephen Hawking also toyed with the idea of a time-reversing universe for a while, only to drop it with the admission that it was his "greatest mistake." Hawking at first believed that applying quantum mechanics to a cyclic universe required detailed time symmetry. It turns out, however, that this is not so—at least, in the standard formulation of quantum mechanics. Recently, the physicists Murray Gell-Mann and James Hartle have discussed a modification to the rules of quantum mechanics, in which the time symmetry is simply *imposed,* and then they have asked whether this state of affairs would have any observable consequences at our cosmic epoch. So far, it is not clear what the answer might be.

A very different way of avoiding cosmic doom has been proposed by the Russian physicist Andrei Linde. It is based on an elaboration of the inflationary-universe theory discussed in chapter 3. In the original inflationary-universe scenario, it was supposed that the quantum state of the very early universe corresponded to a particular excited vacuum that had the effect of temporarily driving runaway expansion. In 1983, Linde suggested that the quantum state of the early universe might instead vary from place to place in a chaotic manner: low energy here, moderately excited there, very excited in some regions. Where the state was excited, there inflation would occur. Furthermore, Linde's calculations of the behavior of the quantum state showed clearly that highly excited states inflate the fastest and decay the slowest, so that the more excited the state was in a particular region of space, the more the universe would inflate in that region. It is clear that after a very short time the regions of space where the energy was accidently greatest, and inflation fastest, would have swelled the most and would occupy the lion's share of the total space. Linde likens the situation to Darwinian evolution, or to economics. A successful quantum fluctuation to a very excited state, although it means borrowing a lot of energy, is immediately rewarded by a huge growth in the volume of that region. So the high-borrowing, superinflating regions soon come to predominate.

As a result of chaotic inflation, the universe would become divided into a cluster of miniuniverses, or bubbles, some inflating like crazy, others not inflating at all. Because some regions—simply as a result of random fluctuations—will have a *very* large excitation energy, there will be much more inflation in those regions than was assumed in the original theory. But because these are precisely the regions to inflate the most, a point selected at random in the post-inflationary universe would be very likely located in such a highly inflated region. Thus our own location in space very probably lies deep within a superinflated region. Linde calculates that such "big bubbles" may have inflated by a factor of 10 to the power 10^8, which is 1 followed by a hundred million zeros!

Our own megadomain would be but one among an infinite number of highly inflated bubbles, so on an enormous scale of size the universe would still look extremely chaotic. Within our bubble—which extends beyond the currently observable universe by a stupendously large distance—matter and energy are distributed approximately uniformly, but beyond our bubble lie other bubbles, as well as regions that are still in the process of inflating. In fact, inflation never ceases in Linde's model: there are always regions of space where inflation is taking place, where new bubbles are forming even as other bubbles pass through their life cycles and die. So this is a form of eternal universe, similar to the baby-universes theory discussed in the last chapter, where life, hope, and universes spring eternal. There is no end to the production of new bubble universes by inflation—and probably no beginning either, although there is currently some contention about that.

Does the existence of other bubbles offer our descendants a lifeline? Can they avoid cosmic doom—or, more accurately, bubbledoom—by always transferring to another, younger bubble in the fullness of time? Linde addressed precisely this question in a heroic paper on "Life after Inflation," published in the journal *Physics Letters* in 1989. "These results imply that life in the inflationary universe will never disappear," he wrote. "Unfortunately, this conclusion does not automatically mean that one can be very optimistic about the future of mankind." Noting that any particular domain, or bubble, will slowly become uninhabitable, Linde concludes: "The only possible strategy of survival which we can see at the moment is to travel from old domains to the new ones."

The discouraging thing in Linde's version of the inflationary theory is that the size of a typical bubble is enormous. He computes that the nearest bubble beyond our own might be so far away that its distance in light-years must be expressed as 1 followed by several million zeros—a number so large it would need an entire encyclopedia of its own to be written out in full! Even at close to the speed of light, it would take a similar number of years to reach another bubble, unless by some extraordinarily good fortune we just happen to be situated near the edge of our bubble. And even this happy circumstance, Linde points out, would obtain only if our universe continues to expand in a predictable manner. The most minute physical effect—one that would be utterly inconspicuous at the present epoch—could eventually determine the way in which the universe expands once the matter and radiation that dominate it at present become infinitely diluted. For example, there could remain in the universe an exceedingly weak relic of the inflationary force that is at present completely swamped by the gravitational effects of matter but which, given the oceans of time needed for beings to escape from our bubble, would eventually make itself felt. In that case, the universe would, after a long enough duration, begin to inflate once more—not in the frenetic manner of the big bang but exceedingly slowly, in a sort of pale imitation of the big bang. However, this feeble whimper, weak though it might be, would continue forever. Although the growth of the universe would accelerate only at a tiny rate, the fact that it accelerates at all has a crucial physical consequence. The effect is to create an event horizon within the bubble, which is rather like a black hole inside out and just as effective a trap. Any surviving beings would become helplessly entombed deep within our bubble, because as they sped toward the edge of the bubble the edge would recede even faster, as a result of the renewed inflation. Linde's calculation, although fanciful, nicely makes the point that the ultimate fate of humankind or our descendants may hinge on physical effects so small that we can have no real hope of detecting them before they start to manifest themselves cosmologically.

Linde's cosmology is in some respects reminiscent of the old steady-state theory of the universe, which was popular in the fifties and early sixties and is still the simplest and most appealing proposal for avoiding the end of the universe. In its original version, expounded by Hermann Bondi and Thomas Gold, the steady-state theory assumed that the universe remains unchanged on a large scale for all time. It therefore has no beginning or end. As the universe expands, new matter is continuously created to fill the gaps and maintain an overall constant density. The fate of any given galaxy is similar to what I have described in the earlier chapters: birth, evolution, and death. But more galaxies are always forming, from the newly created material, which is supplied inexhaustibly. The general aspect of the universe as a whole therefore looks identical from one epoch to the next, with the same total number of galaxies in a given volume of space, consisting of a mixture of various ages.

The concept of a steady-state universe does away with the need to explain how the universe came into existence from nothing in the first place, and it combines interesting variety through evolutionary change with cosmic immortality. In fact, it goes beyond this and provides eternal cosmic youth, because although individual galaxies slowly die, the universe as a whole never grows old. Our descendants never have to grub around scavenging for ever more elusive energy supplies, because the new matter provides it for free. The inhabitants just move on to a younger galaxy when the old one runs out of fuel. And this can continue ad infinitum, with the same level of vigor, diversity, and activity being maintained for all eternity.

There are, however, some physical requirements needed to make the theory work. The universe doubles in volume every few billion years, due to the expansion. To maintain a constant density requires 10^{50} tons or so of new matter to be created over that period. This seems a lot, but on average it amounts to the appearance of just one atom per century in a region of space the size of an aircraft hanger. It is unlikely that we would notice such a phenomenon. A more serious problem concerns the nature of the physical process responsible for creating matter in this theory. At the very least, we should want to know where the energy comes from that supplies the additional mass, and how this miraculous jar of energy manages to be inexhaustible. This problem was tackled by Fred Hoyle, who, with his collaborator Jayant Narlikar, developed the steady-state theory in great detail. They proposed a new type of field—a creation field—to supply the energy. The creation field itself was postulated to have negative energy. The appearance of each new particle of matter with mass m had the effect of contributing an energy $-mc^2$ to the creation field.

Although the creation field provided a technical solution to the problem of creation, it left many questions unexplained. It also seemed rather ad hoc, as no other manifestations of the mysterious field were apparent. More seriously, observational evidence began to mount against the steady-state theory in the 1960s, the most important of which was the discovery of the cosmic background heat radiation. This uniform background receives a ready interpretation as a relic of the big bang, but it is hard to explain convincingly in the steady-state model. In addition, deep-sky surveys of galaxies and radio galaxies showed unmistakable evidence that the universe is evolving on a large scale. When this became clear, Hoyle and his coworkers abandoned the simple version of the steady-state theory, although more complicated variants make fitful reappearances from time to time.

Quite apart from physical and observational problems, the steady-state theory raises some curious philosophical difficulties. For example, if our descendants have truly *infinite* time and resources at their disposal, no obvious limits can be placed on their technological development. They would be free to spread across the universe, gaining control over ever greater volumes of space. Thus a large portion of the universe in the very far future would essentially be technologized. But by hypothesis the large-scale nature of the universe is supposed to be unchanging with time, so the steady-state assumption obliges us to conclude that the universe we see today is *already* technologized. Because the physical conditions in the steady-state universe are overall the same at all epochs, intelligent beings must arise at all epochs too. And because this state of affairs has existed for all eternity, there should be some communities of beings that have been around for an arbitrarily long time and will have expanded to occupy an arbitrarily large volume of space—including our region of the universe—technologizing it. This conclusion is not evaded by supposing that intelligent beings generally have no desire to colonize the universe. It takes only one such community to arise an arbitrarily long time ago for the conclusion to be valid. It is another case of the old conundrum that in an infinite universe anything that is even remotely possible *must* happen sometime, and happen infinitely often. Following the logic to its bitter conclusion, the steady-state theory predicts that the processes of the universe are identical to the technological activities of its inhabitants. What we call nature is, in fact, the activity of a superbeing, or a community of superbeings. This seems like a version of Plato's demiurge (a deity who works within the bounds of physical laws already laid down), and it is interesting that Hoyle, in his later cosmological theories, explicitly advocates such a superbeing.

Any discussion of the end of the universe confronts us with questions of purpose. I have already noted that the prospect of a dying universe convinced Bertrand Russell of the ultimate futility of existence. It is a sentiment echoed in more recent years by Steven Weinberg, whose book *The First Three Minutes* culminates with the stark conclusion that "the more the universe seems comprehensible, the more it also seems pointless." I have argued that the original fear of a slow cosmic heat death was perhaps exaggerated, and may even be erroneous, although sudden death by a big crunch remains a possibility. I have speculated about the activities of superbeings who can achieve miraculous physical and intellectual goals against the odds. I have also looked briefly at the possibility that thoughts may know no bounds, even if the universe does.

But do these alternative scenarios alleviate our sense of unease? A friend of mine once remarked that from what he'd heard of Paradise he wasn't much interested. The prospect of living forever in a state of sublime equilibrium he found utterly unappealing. Better to die quickly and have it all over with than face the boredom of eternal life. If immortality is limited to having the same thoughts and experiences over and over again forever, it does truly seem pointless. However, if immortality is combined with progress, then we can imagine living in a state of perpetual novelty, always learning or doing something new and exciting. The trouble is, what for? When human beings embark on a project for a purpose, they have in mind a specific goal. If the goal is not achieved, the project will have failed (though the experience may not necessarily be valueless). On the other hand, if the goal is attained, the project will be completed and the activity will then cease. Can there be true purpose in a project that is *never* completed? Can existence be meaningful if it consists of an unending journey toward a destination that is never reached?

If there is a purpose to the universe, and it achieves that purpose, then the universe must end, for its continued existence would be gratuitous and pointless. Conversely, if the universe endures forever, it is hard to imagine that there is any ultimate purpose to the universe at all. So cosmic death may be the price that has to be paid for cosmic success. Perhaps the most that we can hope for is that the purpose of the universe becomes known to our descendants before the end of the last three minutes.

Barrow, John D., and Frank J. Tipler, *The Anthropic Cosmological Principle* (Oxford: Oxford University Press, 1986).

Burrows, Adam, "The Birth of Neutron Stars and Black Holes," *Physics Today*, 40 (1987): 28.

Chapman, Clark R., and David Morrison, *Cosmic Catastrophes* (New York & London: Plenum Press, 1989).

Close, Frank, *End: Cosmic Catastrophe and the Fate of the Universe* (New York: Simon & Schuster, 1988).

Coleman, Sidney, and Frank De Luccia, "Gravitational Effects on and of Vacuum Decay," *Physical Review D*, 21 (1980): 3305.

Davies, Paul, *The Cosmic Blueprint* (New York: Simon & Schuster, 1989).

———, *The Mind of God* (New York: Simon & Schuster, 1991).

Dyson, Freeman J., "Time without End: Physics and Biology in an Open Universe," *Reviews of Modern Physics*, 51 (1979): 447.

Gold, Thomas, "The Arrow of Time," *American Journal of Physics*, 30 (1962): 403.

Hawking, Stephen W., *A Brief History of Time: From the Big Bang to Black Holes* (New York: Bantam, 1988).

Hut, Piet, and Martin J. Rees, "How Stable Is Our Vacuum?" *Nature*, 302 (1983): 508.

Islam, Jamal N., *The Ultimate Fate of the Universe* (Cambridge: Cambridge University Press, 1983).

Linde, Andrei D., *Particle Physics and Inflationary Cosmology* (New York: Gordon & Breach, 1991).

Luminet, Jean-Pierre, *Black Holes* (Cambridge: Cambridge University Press, 1992).

Misner, Charles W., Kip S. Thorne, and John A. Wheeler, *Gravitation* (San Francisco: W. H. Freeman, 1970).

Page, Don, and Randall McKee, "Eternity Matters," *Nature*, 291 (1981): 44.

Rees, Martin J. "The Collapse of the Universe: An Eschatological Study," *The Observatory*, 89 (1969): 193.

Smolin, Lee, "Did the Universe Evolve?" *Classical and Quantum Gravity*, 9 (1992): 173.

Tipler, Frank J., *The Physics of Immortality* (New York: Doubleday, 1994).

Tolman, Richard C., *Relativity, Thermodynamics, and Cosmology* (Oxford: Clarendon Press, 1934).

Turner, Michael S., and Frank Wilczek, "Is Our Vacuum Metastable?" *Nature*, 298 (1982): 633.

Waldrop, M. Mitchell, *Complexity: The Emerging Science at the Edge of Order and Chaos* (New York: Simon & Schuster, 1992).

Weinberg, Steven, *The First Three Minutes: A Modern View of the Origin of the Universe*, updated ed. (New York: Basic Books, 1988).